Handbook of Fibers and Polymers

Handbook of Fibers and Polymers

Ajay Kumar Yadav

Delve
PUBLISHING

www.delvepublishing.com

Handbook of Fibers and Polymers

Ajay Kumar Yadav

Delve Publishing
224 Shoreacres Road
Burlington, ON L7L 2H2
Canada
www.delvepublishing.com
Email: orders@arclereducation.com

ISBN: 978-1-77469-134-2 (Hardcover)

ABOUT THE AUTHOR

Ajay Kumar Yadav is presently working as a lecturer at C.S.J.M University Kanpur, India. He has over 10 years of teaching and research experience. He has completed his Ph.D in Pharmaceutical Sciences from SHUATS. He has published several research papers in international and national journals and has also presented many posters.

TABLE OF CONTENTS

List of Figures .... *xi*
List of Tables .... *xvii*
Abbreviations .... *xix*
Preface .... *xxiii*

Chapter 1 **Introduction to Fiber Materials 1**
1.1. Introduction 2
1.2. Flexibility 5
1.3. Fiber Spinning Processes 7
1.4. Stretching and Orientation 10
1.5. Types and Properties of Fibers 10
1.6. Waste Materials and Fibers 31
1.7. Important Factors to Consider When Using Fibers 34
References 35

Chapter 2 **Fundamentals of Polymers 47**
2.1. Introduction 48
2.2. The History of the Concept of the Macromolecule 49
2.3. Classification of Polymers 51
2.4. Structure and Properties of Polymers 53
2.5. Thermoplastic Polymers 53
2.6. Thermosetting Polymers 61
2.7. Naturally Occurring Polymers 68
References 73

Chapter 3 **Introduction of Fiber-Reinforced Polymers 81**
3.1. Introduction 82
3.2. Basic Concepts and Manufacturing Process 86
3.3. Types of Polymer Composites 88

3.4. Fiber Manufacturing Process ... 93
3.5. Composite Molding ... 96
3.6. Resin Infusion ... 101
3.7. Glass Fiber Material ... 103
3.8. Aramid Fiber Material Process ... 104
3.9. FRP Applications ... 104
3.10. Mechanical Properties Measurements ... 107
3.11. Real Polymer's Mechanical Properties ... 112
References ... 113

Chapter 4 Glass Fiber-Reinforced Polymers ... 125
4.1. Introduction ... 126
4.2. Chemical Composition and Classification ... 127
4.3. Fabrication of the Glass Fibers ... 130
4.4. Forms of the Glass Fibers ... 132
4.5. Properties of Glass Fiber ... 135
4.6. Glass Fibers in the Polymer Composites ... 139
4.7. Applications ... 144
References ... 150

Chapter 5 Properties, Fabrication, and Applications of Carbon Fiber-Reinforced Polymers ... 159
5.1. Introduction ... 160
5.2. Manufacturing Processes ... 161
5.3. Surface Treatment ... 162
5.4. Materials for CFRP ... 165
5.5. Surface Treatment of the Carbon Fibers ... 166
5.6. Preparation of Carbon Fiber-Strengthened Polymer Composites ... 170
5.7. Characterization of the Carbon Fibers ... 170
5.8. Characterization of CFRPCS ... 171
5.9. Electrochemical Oxidation Effect ... 176
5.10. Electroplating Effect ... 179
5.11. Applications ... 186
References ... 189

Chapter 6 Kevlar Fiber-Reinforced Polymer Composites 197
6.1. Introduction 198
6.2. Kevlar Fiber 199
6.3. Interface 201
6.4. Factors Affecting the Composite Properties 202
6.5. Surface Modification 207
References 211

Chapter 7 Nylon Fiber-Reinforced Polymeric Materials 217
7.1. Introduction 218
7.2. Nylon Fibers Used as Reinforcements 219
7.3. Matrices and Applications 223
7.4. Manufacturing af Nylon-Reinforced Composites 227
References 232

Chapter 8 Polyester Fiber-Reinforced Polymeric Materials 237
8.1. Introduction 238
8.2. Synthesis and Basic Properties of Polyester Fibers 240
8.3. Polyester Fiber-Reinforced Polymer Composites 246
References 251

Index 255

LIST OF FIGURES

Figure 1.1. Samples of steel and copper fibers (top row), and basalt, plastic, and coconut fibers (bottom row)

Figure 1.2. Decrease in strength (sf) of a carbon fiber with increase in diameter (Lamotte and Perry, 1970)

Figure 1.3. Schematic illustration of wet-spinning process

Figure 1.4. Fiber diameter of different materials with flexibility equal to that of a nylon fiber of diameter equal to 25 mm. Note that one can make very flexible fibers out of brittle materials such as glass, silicon carbide, alumina, etc., provided one can process them into a small diameter

Figure 1.5. Melt-spinning process

Figure 1.6. Schematic diagram of dry-jet wet spinning technique

Figure 1.7. Different categories of natural fibers

Figure 1.8. Cross-sectional view of bast fiber

Figure 1.9. Schematic illustration of fiber extraction from date-palm tree

Figure 1.10. Different crockery items manufactured from wheat straw fibers

Figure 1.11. Different types of steel fibers

Figure 1.12. Different types of glass fibers

Figure 1.13. Different types of carbon fibers

Figure 1.14. Applications of aramid fibers

Figure 1.15. Typical processing steps involved in the synthesis of a ceramic fiber

Figure 1.16. Different types of basalt fibers

Figure 2.1. Classification of polymers

Figure 2.2. Different types of macromolecules

Figure 2.3. Detailed classification of polymers

Figure 2.4. The structural formula of polythene

Figure 2.5. The structural formula of polypropylene

Figure 2.6. The structural formula of PMMA polymer

Figure 2.7. Structure of vinyl benzene (styrene)

Figure 2.8. Structural formula of PVC

Figure 2.9. The structural formula of epoxy

Figure 2.10. The segment of the phenol-formaldehyde polymer

Figure 2.11. Structures of urea and melamine

Figure 2.12. Polytetrafluoroethylene structure

Figure 2.13. The structural formula of cellulose

Figure 2.14. The structural formula of starch

Figure 2.15. Structures of protein: (a) primary structure; (b) secondary structure; (c) tertiary structure; (d) quaternary structure

Figure 3.1. Fiber-reinforced polymer

Figure 3.2. Applications in which carbon fiber reinforced composites are used

Figure 3.3. Different manufacturing processes of composite

Figure 3.4. The impact strength comparison of polymer composites (at room temperature) reinforced by silk fibers, glass fibers, plant fibers, and carbon fibers

Figure 3.5. Different raw materials used to synthesize the biocomposites

Figure 3.6. Fiber preforms: (a) a commercial 3D printer used to print a hollow-core fiber preform; (b) high-temperature pre-caning for reducing the size of the preform followed by; (c) cane drawing; and (d) fiber drawing

Figure 3.7. Bladder assisted composite manufacturing (BACM) fabrication assembly along with the process schematic

Figure 3.8. Compression molding illustrated schematically

Figure 3.9. Vacuum bagging autoclave technique

Figure 3.10. The composite manufacturing process of the mandrel wrapping

Figure 3.11. Filament winding process

Figure 3.12. The pultrusion process illustrated schematically

Figure 3.13. Vacuum resin infusion process

Figure 3.14. The polymer material's intrinsic deformation behavior schematically represented

Figure 3.15. Stress-strain curve for a material that has varying toughness and tensile strength

Figure 3.16. A stress-strain plot of tough and strong polymers

Figure 4.1. Piezoelectric elements incorporated into the module containing glass fiber reinforced plastics (GFRP)

Figure 4.2. Representation of glassy silica's chemical network

Figure 4.3. Diagram of the production process of glass fiber

Figure 4.4. Images of diverse forms of the glass fibers: (a) roving on a coil; (b) woven fabric; (c) braided fabric; (d) chopped strand mat

Figure 4.5. Cross-sections of the GFRP sample: (a) optical microscopy; and (b) computed tomography

Figure 4.6. Utilization of the composite materials in present aeronautic structures: EC-135 helicopter, Airbus A380 airliner, and Eurofighter jet

Figure 4.7. Boat made from GFRP

Figure 4.8. Diverse body parts made from GFRP

Figure 4.9. GFRP based tanks and pipes

Figure 4.10. GFRP based blades of wind turbine

Figure 5.1. The probable functional groups introduced on the oxidized surfaces of carbon

Figure 5.2. Schematic representation of constant electrolytic surface anodization procedure

Figure 5.3. Schematic representation of the oxyfluorination reactor: (1) F_2 (fluorine) gas cylinder; (2) N_2 (nitrogen) gas cylinder; (3) O_2 (oxygen) gas cylinder; (4) the buffer tank; (5) HF absorber; (6) reactor; (7) pressure gauge; (8) fluorine absorber (aluminum oxide); (9) glass cock; (10) nitrogen in liquid form; and (11) the rotary vacuum pump

Figure 5.4. Schematic representation of the atmospheric pressure plasma devices

Figure 5.5. Schematic representation of the principles of capillary rise technique

Figure 5.6. Usual data of wicking rate in weight m^2 as the function of time (t)

Figure 5.7. Schematic representation of SEN (single-edge notched) beam fracture test of toughness

Figure 5.8. Schematic representation of DCB (double cantilever beam) specimen

Figure 5.9. Load-extension curves produced by three key classes of the fracture behaviors

Figure 5.10. Evolution of K_{IC} as the function of current density planned

Figure 5.11. Dependence of the K_{IC} of composites on a specific (or polar) component of a fiber surface free energy planned

Figure 5.12. XPS spectra of untreated and Ni-plated carbon fibers: (a) 0 A m^{-2}; (b) 5 A m^{-2}; (c) 10 A m^{-2}; (d) 30 A m^{-2}; and (e) 60 A m^{-2}

Figure 5.13. High-resolution Ni_{2p} and O_{1s} XPS spectra of the Ni-plated carbon fibers (10 A m^{-2} current density). (a) NiO (peak 1, BE= 529.6 eV), C=O or OH (peak 2, BE= 531.6 eV), OCO (peak 3, BE= 532.6 eV; (b) Ni-metal (peak 1, BE= 852.7 eV), NiO (peak 2, BE= 854.6 eV), Ni(OH)2 (peak 3, BE= 856.4 eV), and Ni-metal (peak 4, BE= 858.5 eV)

Figure 5.14. Free energy of surface the Ni-plated and untreated carbon fibers [gS: surface free energy; gLS: London dispersive component; and gSPS: specific component]

Figure 5.15. KIC of nickel-plated carbon fiber-strengthened composites with diverse current densities

Figure 5.16. Dependence of K_{IC} on the O_{1s}/C_{1s} ratio (where R → coefficient of regression)

Figure 5.17. Dependence of K_{IC} on the specific polar component $g^{SP}{}_{S}$ of the free energy of surface (where R → coefficient of regression)

Figure 5.18. XPS broad scan spectra of oxyfluorinated carbon fibers

Figure 6.1. Applications of aramid fiber

Figure 6.2. Molecular structure of Kevlar and hydrogen bonding

Figure 6.3. Illustrations of a classic fracture surface of unidirectional composite loaded in tension with the fiber direction: (a) composite for strong interfacial bond; (b) composite for an intermediate interfacial bond; and (c) composite for a poor interfacial bond

Figure 6.4. Diagrammatic representation of: (a) aligned and continuous; and (b) randomly oriented and discontinuous fiber-reinforced composites

Figure 6.5. Flow regions along the injection molding process

Figure 6.6. Surface modification of aramid fibers

Figure 7.1. Rod-shaped Si_3N_4 particle involving a nylon 6 matrix

Figure 7.2. Optical micrograph of a chain-shaped fiber

Figure 7.3. Pull-out load versus displacement curves: (a) straight fibers; and (b) chain-shaped fibers

Figure 7.4. FESEM micrographs of: (a) the cross-section of a nylon 6 electrospun fiber mat; and (b) of a detail of the electrospun fibers

Figure 7.5. SEM micrographs at a various magnification of the cryogenic fracture surface of a sample of PCL involving nylon 6 electrospun fibers

Figure 7.6. Optical micrograph of a PMMA/nylon composite formed by *in-situ* polymerization in supercritical CO_2

Figure 7.7. Fracture morphology of nylon 6/6 composite, and (f) PMMA/2.5% nylon 6 fiber-reinforced PMMA composites: (a) pure composite (tensile fracture surface) Reprinted PMMA; (b) PMMA/0.5% nylon 6 composite; (c) PMMA/1.5% nylon 6 composite; (d) PMMA/2.5% nylon 6 composite; (e) PMMA/3.5% nylon

Figure 7.8. SEM micrograph of nylon fibrils achieved *in-situ* in a PET matrix

Figure 8.1. Chemical structure of PET

Figure 8.2. Polymer fiber melt spinning production process

Figure 8.3. Tensile properties of PET fibers. (A) high tenacity filament; (B) high tenacity staple; (C) regular tenacity filament; (D) regular tenacity staple; (E) POY thread

Figure 8.4. Surfaces of compression-molded PET/PP/E-GMA mixtures with an MFC structure in different ratios (by wt.%). (a) 40/60/0; and (b) 40/54/6. PP was selectively dissolved leaving PET fibrils are visible

Figure 8.5. Tensile properties of PET/PP microfibrillar composites

Figure 8.6. A fiber of polyester pulled out from PP matrix displaying characteristic kink bands

LIST OF TABLES

Table 1.1. Differentiating micro synthetic fibers with macro synthetic fibers

Table 1.2. Classification of basalt

Table 1.3. Difference between short and continuous basalt fiber

Table 2.1. Details of processes for preparing poly(ethylene)

Table 4.1. Chemical composition of different types of glass fiber in weight percent

Table 4.2. Properties of the commercial glass fibers

Table 5.1. Distinctive properties of three grades of the carbon fibers

Table 5.2. Description of some carbon fibers

Table 5.3. Operating conditions and composition of a nickel-plating bath

Table 5.4. Experimental oxyfluorinated conditions of the PAN-centered carbon fibers utilized

Table 5.5. Features of the wetting liquids utilized in this work

Table 5.6. Variation of the surface chemical compositions of carbon fibers with the current density perceived by XPS measurement (at percent)

Table 5.7. Fluctuation of the functional groups of the surfaces of carbon fibers with the current density perceived from XPS measurement (at percent)

Table 5.8. Determinations of contact angle determinations on carbon fibers as the function of current density

Table 5.9. Outcomes of the free energy of surface and their constituents of carbon fibers as the function of current density

Table 5.10. O_{1s}/C_{1s} ratio and elemental composition of the Ni-plated and untreated carbon fibers (at percent)

Table 6.1. Unique properties of Kevlar fiber in comparison with other fibers

Table 7.1. Representative properties of various reinforcement fibers

Table 8.1. PET market size per year

Table 8.2. Physical properties of polyester fibers

LIST OF ABBREVIATIONS

ABS	Acrylonitrile Butadiene Styrene
Al_2O_3	Aluminum Oxide
B_2O_3	Boric Oxide
BMC	Bulk Molding Compound
CaO	Calcium Oxide
CFRCs	Carbon-Fiber-Reinforced Polymer Composites
CH_2CHOH	Polyvinyl Alcohol
CSM	Chopped Strand Mats
CTE	Coefficient of Thermal Expansion
DCB	Double Cantilever Beam
DDM	Diamino Diphenylmethane
DGEBA	Diglycidylether of Bisphenol-A
DMT	Dimethyl Terephthalate
EG	Ethylene Glycol
EP	Epoxy
F	Force
Fe_2O_3	Ferric Oxide
FeO	Ferrous Oxide
FRP	Fiber-Reinforced Plastic
FTC	Federal Trade Commission
GFRP	Glass-Fiber-Reinforced Plastics
GMT	Glass Mat Thermoplastic
GPa	Gigapascals
HDPE	High-Density Polyethylene
IFSS	Interfacial Shear Strength

ILSS	Interlaminar Shear Strength
LEFM	Linear Elastic Fracture Mechanics
MEK	Methyl Ethyl Ketone
MFC	Mass Flow Controller
MFC	Microfibrillar-Reinforced Composite
MFR	Microfibrillar-Reinforced
MgO	Magnesium Oxide
MPa	Megapascals
NBR	Acrylonitrile Butadiene Rubber
Ni	Nickel
NR	Natural Rubber
O_2	Oxygen
ORNL	Oak Ridge National Laboratory
PA	Polyamide
PAN	Polyacrylonitrile
PBT	Polybutylene Terephthalate
PC	Polycarbonate
PE	Polyethylene
PEEK	polyether Ether Ketone
PEN	Poly(Ethylene Naphthalate)
PET	Poly(Ethylene Terephthalate)
PF	Phenolic
PI	Polyimide
PMMA	Polymethylmethacrylate
POM	Polyoxymethylene
POY	Partially Oriented Yarn
PP	Polypropylene
PPTA	Poly(p-Phenylene Terephthalamide)
PS	Polystyrene
PTA	Purified Terephthalic Acid
PUR	Polyurethane
PVC	Polyvinyl Chloride
RTM	Resin Transfer Molding

SEN	Single-Edge Notched
SiO_2	Silicon Dioxide
SMC	Sheet Molding Compound
UD	Unidirectional
VARTM	Vacuum Assisted Resin Transfer Molding
VE	Vinyl Ester
WPC	Wood-Plastic Composites
XPS	X-Ray Photoelectron Spectroscopy

PREFACE

This book contains a comprehensive collection of data on different types of fibers, polymer matrices, and composite structures employed to develop a wide range of materials from polymers and natural fibers.

The prospect of achieving any specific set of properties has stimulated the wide-ranging applications of composite materials in numerous industries. Composites, nanocomposites, and biocomposites are being recognized as a superior class of materials for aerospace, sports, transportation, mining, civil, electronics, and many other industrial sectors. Fibrous materials have been broadly used in composites in various forms (tows, fabrics, nanofiber webs, sort fibers, mats, etc.), for strengthening purposes or for adding several functionalities. Vast flexibility of structure designing, material selection, and attainable features led to their widespread use in many industrial applications.

In recent years, high-stiffness, high-strength, and lightweight composite materials have gained recognition as a significant cutting-edge technology with a potential to generate efficient power, reinvent the high-efficiency transportation sector, offer new carriages for storage and transportation of reduced carbon fuels, and enhance renewable energy production. It is imperative to harness the true potential of composite materials by utilizing advanced engineering techniques to manufacture cost-effective yet efficient products at a commercial scale by utilizing different combinations of polymers and fibers.

This book discusses various categories of fibers and polymers for the construction of composite materials for different applications. A preliminary discussion on the fundamental characteristics of fiber materials and their manufacturing techniques is offered in Chapter 1. Chapter 2 focuses on the fundamental properties of polymeric materials and their potential as a matrix material in composites. Chapter 3 comprehensively discusses the key concepts regarding the development of composite materials from different combinations of fibers and matrices. Chapter 4 provides information about properties, manufacturing techniques, and applications of glass fiber reinforced composites.

Chapter 5 offers a detailed analysis of fundamental properties, synthesis techniques, and applications of carbon fiber-reinforced composites. Chapter 5 offers a detailed analysis of fundamental properties, synthesis techniques, and applications of carbon fiber-reinforced composites.

Therefore, it is believed that this book can serve as a greatly valuable reference for a varied range of readers, including scientists, engineers, policymakers, and professors. This book is also useful for teachers and students from different backgrounds, including textile engineering, materials science, fiber science and technology, mechanical engineering, nanotechnology, chemical engineering, medical sciences, and environmental science.

In the end, I would like to extend my sincere gratitude to the members of modern scientific society for their kind support and help throughout the compilation of the book.

Chapter 1

Introduction to Fiber Materials

CONTENTS

1.1. Introduction 2

1.2. Flexibility 5

1.3. Fiber Spinning Processes 7

1.4. Stretching and Orientation 10

1.5. Types and Properties of Fibers 10

1.6. Waste Materials and Fibers 31

1.7. Important Factors to Consider When Using Fibers 34

References 35

1.1. INTRODUCTION

Fibers HAVE been utilized for building different structures for many years. Through foregoing research and surveys, the usage of natural and artificial fibers has revealed favorable results, as their existence has validated substantial benefits in terms of the whole physical and mechanical properties of the composite material. While comparing fiber reinforcement to traditional reinforcement, the ratio of fiber required is considerably less, creating fiber reinforcement both energy and economically proficient. In recent times, waste fibers have been considered for their prospective as reinforcement in several structural applications. An increase in the number of waste materials in the entire world is an identified issue, as landfill space is inadequate, and the burning process entails substantial energy and yields undesirable emissions. The consumption of waste fibers in construction and industrial materials can relieve these problems and uphold eco-friendly and sustainable solutions that work in the industry.

People employ materials to build and make structural and nonstructural products in a variety of industries (Siddika et al., 2019). Materials are essential to produce bricks, and bricks can be utilized to construct houses, which are a basic need for a community. Earth materials are still being extensively used in various industrial applications even after thousands of years (Laborel-Préneron et al., 2016). Indeed, the types of industrial material differ all over the world, as economies and readily available materials change.

Prolonged history of composite materials exists in which fibers have been used as reinforcement to improve the properties of the materials (Jamshaid and Mishra, 2016). Seeing back to the Egyptian era, natural fibers like straw and horsehair were integrated into the development of mud bricks. Straw was a common material used in initial Japanese and Chinese construction, purposed as reinforcement for structural components of houses. As of the late 1800s, the United States implemented straw material as a key component of bearing walls (Hejazi et al., 2012).

Specifically, coconut (coir), sisal, straw, palm, flax, jute, bamboo, and can have an extended history as they are readily accessible and can be acquired from the natural environment. With natural fibers, their physical and mechanical properties will be affected by the surrounding environment at the time of harvest. This comprises the condition of the soil, the process employed for the extraction, any treatment commenced, temperature, air humidity, and more (Ghavami et al., 1999). Concrete is a common component utilized in the construction industry. For about 50 years, fibers have been incorporated into concrete applications in the construction and civil industry (Altoubat et al., 2008). In recent times, fiber-reinforced polymers have been employed in the construction industry for over two decades, representing positive attributes such as increased strength and being lightweight and resistant to corrosion (Myadaraboina et al., 2014).

Every fiber material has reduced as well as high properties depending upon the type of properties. Such as, a significant property for which concrete is recognized is its extraordinary compressive strength. On the contrary, it holds low tensile strength. Similarly, the soil is usually poor in tensile and shear strength. Particular fibers can be selected and incorporated into the prevailing materials to increase the properties in which the parent material is rundown, creating the resultant material well-rounded. In the case of synthetic fibers, diverse production procedures and raw material configurations will initiate different mechanical properties of the fibers (Behfarnia and Behravan, 2014; Yin et al., 2015).

Fibers have different features and come in a variety of shapes and sizes. If employed, the fiber category will need to be cautiously deliberated and chosen, as various fibers, such as glass, basalt, and recycled polyethylene (PE) terephthalate fibers, can be damaged in alkaline environments (Roesler et al., 2006; Pelisser et al., 2012).

Figure 1.1. Samples of steel and copper fibers (top row), and basalt, plastic, and coconut fibers (bottom row).

Source: https://www.mdpi.com/1996-1944/12/16/2513.

Fibers can lessen the aggregate expense of the manufacturing, as they can be utilized as reinforcement, substituting the traditional and energy-consuming procedures of wired mesh and steel reinforcement bars. Consequently, there will be a reduced labor cost, low preservation cost, less construction time, and lower manufacturing cost. Energy will also be saved, as the amount of fiber utilized is usually considerably less than the volume of raw material required to fabricate the traditional reinforcement (Figure 1.1) (Lloyd, 2014).

Reinforcements need not essentially be in the form of long fibers. These also exist in the form of particles, whiskers, short fibers, flakes, continuous fibers, or sheets. It has been observed that most reinforcements incorporated in composites have a fibrous nature because materials are stronger and stiffer in the fibrous form as compared to other forms. Explicitly, in this classification, we are utmost attracted to the so-called advanced fibers, which hold very high strength and very high stiffness with reduced density. The reader should apprehend that numerous naturally arising fibers can be and are utilized in conditions concerning not very high stresses (Chawla and Bastos, 1979; Chawla, 2012). The pronounced advantage in this case, certainly, is its reduced expense. The vegetable kingdom is, indeed, the major source of fibrous materials. Cellulosic fibers in the form of cotton, hemp, flax, jute, sisal, and ramie, for instance, have been used in the textile industry, whereas wood and straw have been consumed in the paper industry. Other natural fibers, such as hair, wool, and silk, comprise different forms of protein. Silk fibers created by a variety of spiders, specifically, seem to be very attractive despite their high work of fracture. Any discussion of such fibers is beyond the scope of this book. The interested reader is directed to some books that cover the vast field of fibers used as reinforcements (Chawla, 1981; Warner, 1995). In this chapter, we restrain ourselves to different manmade reinforcements. Glass fiber, in its different forms, has been the most used reinforcement for polymer matrices. Aramid fiber, introduced in the 1960s, is considerably stiffer and lighter than glass fiber. Kevlar is Du Pont's trade name for aramid fiber while Twaron is the trade name of aramid fiber prepared by Teijin Aramid. Gel-spun PE fiber, having stiffness comparable to that of aramid fiber, was commercialized in the 1980s. Other high-performance fibers retaining high strength with high stiffness are boron, carbon, silicon carbide, and alumina. These were all developed in the second part of the 20th century. Particularly, some ceramic fibers were manufactured in the last quarter of the 20th century using novel processing techniques, namely, sol-gel processing and controlled pyrolysis of organic precursors.

The usage of fibers as high-performance engineering materials is based on three significant characteristics (Dresher, 1969):

- A small diameter about its grain size or another microstructural unit. This permits a higher segment of the theoretical strength to be achieved than is possible in a bulk form. This is a direct result of the purported size effect; the smaller the size, the lesser the probability of having deficiencies in the material. Figure 1.2 displays that the strength of a carbon fiber drops as its diameter rises (de Lamotte and Perry, 1970). Though this figure indicates a linear drop in strength with increasing fiber diameter, a nonlinear relationship is not exceptional. Figure 1.2 should be taken only as a common trend display.
- A high aspect ratio (length/diameter, l/d), which permits a very large portion of the applied load to be transmitted via the matrix to the stiff and strong fiber.
- A very high degree of flexibility, which is a distinguishing property of a material that has a low modulus or stiffness and a small diameter. This flexibility permits the usage of a variety of techniques for manufacturing composites with these fibers.

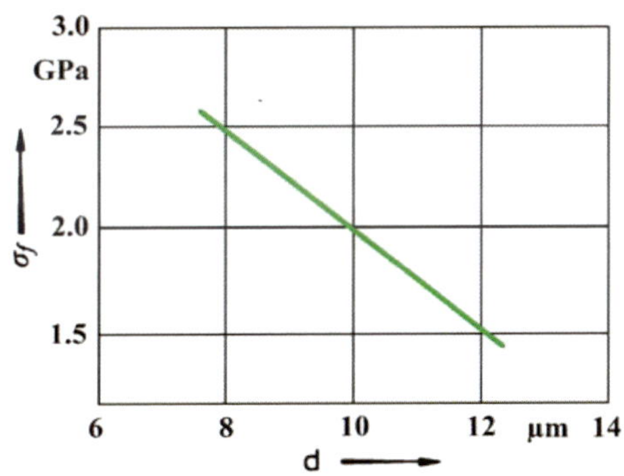

Figure 1.2. Decrease in strength (sf) of a carbon fiber with increase in diameter (Lamotte and Perry, 1970).

Source: https://www.sciencedirect.com/science/article/abs/pii/0015056870900217.

Next, we deliberate the perception of flexibility, and then we define the general fiber spinning procedures.

1.2. FLEXIBILITY

The flexibility of a certain material, as indicated in the characteristics above, is a function of its elastic stiffness and dimensions of the cross-section.

Instinctively, one can certainly envisage that the higher the stiffness of the material, the low flexible it will be. Nevertheless, when we deliberate flexibility of fiber or thread, we desire to identify what arbitrary radius we can bend it before it fails. We can assume our single fiber to be an elongated elastic beam. Let us subject this fiber of Young's modulus, E, and diameter, d to a bending moment, M, which will bend it to a radius, R. For such an elastic bending of a beam, we describe flexural rigidity as MR. From the elementary strength of materials, we have the subsequent relationship for a beam bent to a radius R:

$M/I = E/R$

$MR = EI$

where; E is Young's modulus of the material, and I is the second moment of area or moment of inertia of its cross-section. For a beam or fiber of diameter, d, the second moment of area about an axis through the centroid of the beam is given by $I = \pi d^4/64$. Now, the flexibility of the fiber (i.e., the elastic beam under consideration) can be defined as the inverse of flexural rigidity. In other words, the flexibility of fiber is an inverse function of its elastic modulus, E, and the second moment of area or moment of inertia of its cross-section, I. The elastic modulus of a material is usually independent of its form or size and is termed as a material constant for a particular chemical composition (supposing a fully dense material). Accordingly, for a specified composition and density, the flexibility of a material is determined by its shape, or more accurately by its diameter. Substituting for $I = \pi d^4/64$ in the above expression, we acquire:

$$MR = EI = \frac{E\pi d^4}{64},$$

or, the flexibility, being equal to 1/MR, is:

$$\text{Flexibility} = \frac{1}{MR} = \frac{64}{E\pi d^4}, \qquad (1)$$

(1)

where; d is the equivalent diameter; and I is the moment of inertia of the beam (fiber). Eqn. (1) specifies that flexibility, 1/MR, is a very profound function of diameter, d. We can summarize the important insinuations of Eqn. (1) as follows:

- The flexibility of fiber is a very critical inverse function of fiber diameter, d.
- At an adequately small diameter, it is conceivable to yield a fiber as flexible as any from a polymer, a ceramic, or a metal.

- Highly flexible fibers can be made out of characteristically brittle materials like glass, silicon carbide, alumina, etc., only if one can shape these brittle materials into a fine diameter fiber. Generating a fine diameter ceramic fiber, though, is an intimidating problem in processing.

To demonstrate this conception of flexibility, we plot the diameter of different materials in fibrous form with the flexibility (1/MR) equivalent to that of exceedingly flexible filament, named as 25-mm-diameter nylon fiber as a function of the elastic modulus, E. Further, given an adequately small diameter, it is probable for metal or ceramic to have a similar degree of flexibility as that of a 25-mm-diameter nylon fiber; it is, though, the additional matter that attaining such a small diameter in practice can be excessively costly (Mohajerani et al., 2018).

1.3. FIBER SPINNING PROCESSES

Fiber spinning is the procedure of extruding a liquid via tiny holes in a spinneret to produce solid filaments. Naturally, silkworms and spiders produce continuous filaments by this method. There occurs a wide range of diverse fiber spinning techniques; some of them are given in subsections.

1.3.1. Wet Spinning

Extrusion of the solution is carried out into a coagulating bath. The jets of liquid freeze or harden in the coagulating bath because of chemical or physical changes (Figure 1.3).

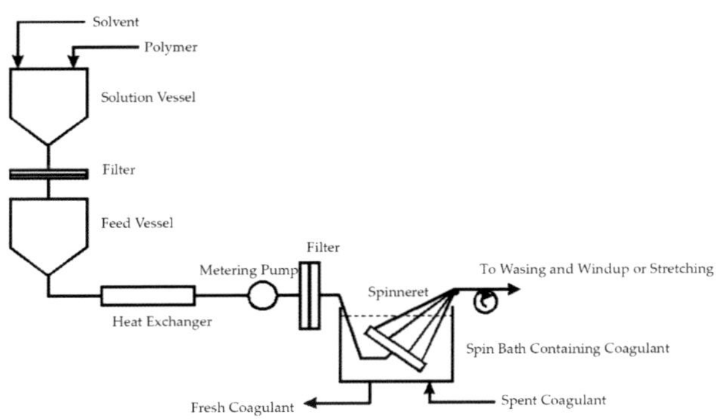

Figure 1.3. Schematic illustration of wet-spinning process.

Source: https://www.pslc.ws/macrog/mpm/composit/fiber/process/wetspin.htm.

1.3.2. Dry Spinning

A solution comprising a fiber-forming material and a solvent is extruded via a spinneret. A stream of hot air intrudes on the jets of solution evolving from the spinneret, the solvent evaporates, thereby producing solid filaments (Figure 1.4).

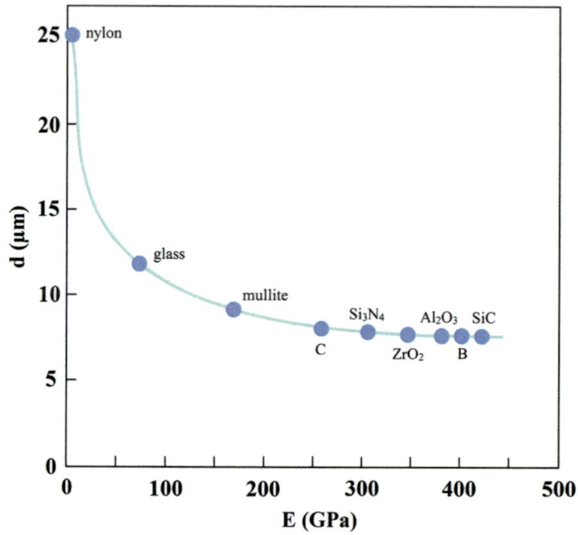

Figure 1.4. Fiber diameter of different materials with flexibility equal to that of a nylon fiber of diameter equal to 25 mm. Note that one can make very flexible fibers out of brittle materials such as glass, silicon carbide, alumina, etc., provided one can process them into a small diameter.

Source: https://link.springer.com/chapter/10.1007/978-0-387-74365-3_2.

1.3.3. Melt Spinning

The fiber-forming material is converted into molten form by heating beyond its melting point, and later, the molten material is extruded through a spinneret. The liquid jets are hardened into solid filaments because of air emerging out from the tiny spinneret holes (Figure 1.5).

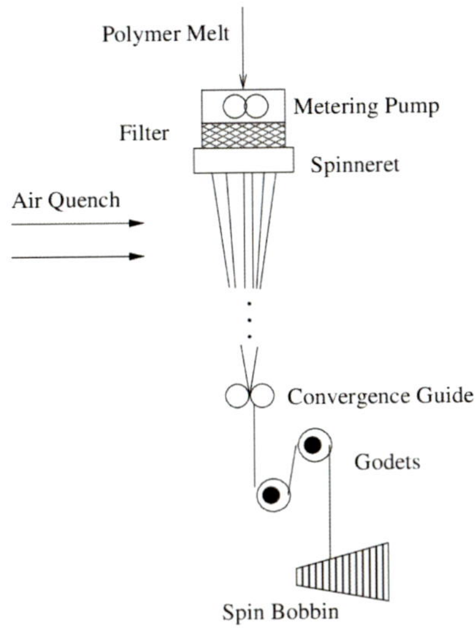

Figure 1.5. *Melt-spinning process.*

Source: https://www.researchgate.net/figure/Fiber-Melt-Spinning-Process_fig3_228922758.

1.3.4. Dry Jet-Wet Spinning

This is a distinctive method developed for the spinning of aramid fibers. In this route, a suitable polymer liquid crystal solution is extruded through spinneret holes, passes through an air gap before arriving at coagulation bath, and at that point goes on a spool for winding purposes (Figure 1.6).

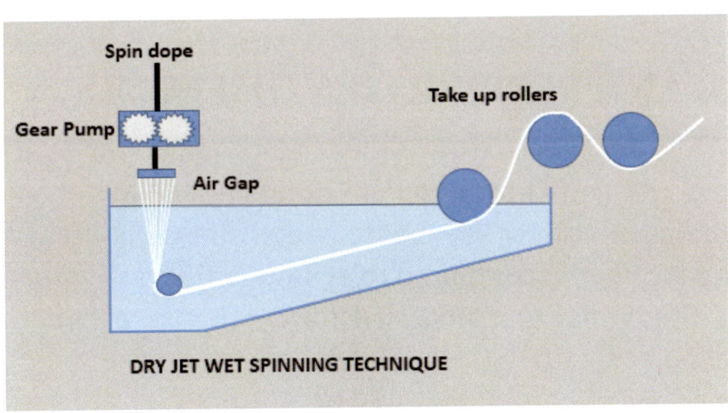

Figure 1.6. *Schematic diagram of dry-jet wet spinning technique.*

Source: https://www.nal.res.in/en/techniques/dry-jet-wet-fiber-spinning-technique.

1.4. STRETCHING AND ORIENTATION

The process of extrusion using a spinneret outcome in a certain chain orientation in the filament. Normally, the molecules present in the surface region endure more orientation compared to molecules located in the interior as the edges of the spinneret hole influence the near-surface molecules more. This effect is termed the skin effect, and it can affect several other characteristics of the fiber, like a bond with a polymeric matrix or the capability to be dyed. Usually, the as-spun fiber is imperiled to certain stretching, initiating further chain orientation along the fiber axis and subsequently better tensile properties, like stiffness and strength, along the fiber axis (De Lamotte and Perry, 1970). The extent of stretch is normally given in terms of a draw ratio, which is the ratio between the initial diameter and the final diameter. For instance, nylon fibers are normally subjected to a draw ratio of 5 after spinning. A high draw ratio outcome in a high elastic modulus. Improved alignment of chains means a higher degree of crystallinity in a fiber. This also determines the capacity of a fiber to absorb moisture. The higher the degree of crystallinity, the lower the moisture absorption capability of fiber. Generally, the higher degree of crystallinity interprets into greater resistance to penetration by foreign molecules, i.e., better chemical stability. The stretching treatment assists to orientate the molecular structure along the fiber axis. It does not, mostly, result in thorough eradication of molecular branching; that is, one gets molecular orientation but not extension. Such stretching treatments surely bring about somewhat more efficient packing than in the unstretched polymer, but there exists a limit to the amount of stretch that can be provided to a polymer as the phenomenon of necking can intercede and result in rupture of the fiber.

1.5. TYPES AND PROPERTIES OF FIBERS

Fibers are usually categorized as either natural fibers or synthetic fibers. Natural fibers are obtained from the environment and are characteristically extracted from the exterior of plants, trees, and straw. Natural fibers possess an extended history while compared to synthetic fibers. Synthetic fibers are manufactured fibers that are produced for a purpose depending upon the application (Saha et al., 2016).

1.5.1. Natural Fibers

Natural fibers include bamboo, coconut (coir), jute, flax, sisal, palm, straw, cane, and many more. Moreover, natural fibers can be classified into various subcategories, like cereal straw, wood fibers, palm tree fibers, basalt fibers, and leaf fibers. With natural fibers, the environmental condition at the time of harvest will have an impact on the physical and mechanical properties. This comprises the condition of the soil, the extraction process, air humidity, temperature, any treatment undertaken, and more (Figure 1.7) (Boukhattem et al., 2017).

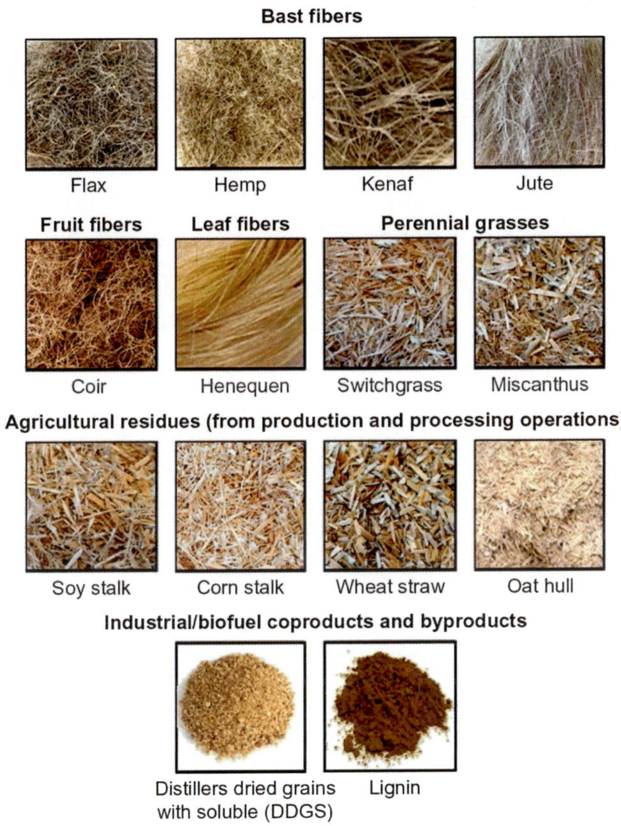

Figure 1.7. Different categories of natural fibers.

Source: https://www.sciencedirect.com/science/article/pii/B9781782423737000019.

The consumption of natural fibers devours a prolonged history. Various natural fibers are found all over the world and are easily obtainable as they cultivate naturally in the environment. They are readily available, abundant, need very little to no energy to produce, cost-effective, can be extracted from waste material, and can decrease the environmental impact when used in the various industries. The usage of natural fibers can also assist the utilization of sustainable materials. Though, it should be eminent that the foremost drawback of these fibers is that, typically, they possess reduced strength, and as they are natural, they possibly will degrade by natural means over time. One more disadvantage of plant fibers is that these fibers are hydrophilic (Kadir et al., 2016).

Some recent studies have tried to enhance the durability of the natural fibers for composite materials. These comprise a coating, chemical treatment, and substituting or blocking the hydroxyl group of the organic natural fibers. However, further research is obligatory to regulate the improvement in durability. It has been indicated that these approaches are quite costly and could be harmful as leeching and exposure to chemical, thermal, biological, and UV-related degradation have not been explored in detail.

A quantity of natural fibers is agricultural waste yields that either get discarded in landfills or burnt. Fiber-reinforced composites containing natural fibers demonstrate the potential for recyclability and biodegradability (Bouasker et al., 2014). Natural fiber-reinforced composites can possess better properties, like enhanced strength, and control shrinkage, and the creation of flaws. By utilizing natural fibers, waste materials in landfills can be minimized, emissions can be avoided, and the energy consumed to burn the waste can be saved.

Readily renewable materials with optimum properties and ecology have been an extensive study in the civil engineering industry. Two foremost techniques to attain sustainability are to use natural fibers and industrial waste materials. Natural materials tend to have better thermal insulation properties and can improve the regulation of temperature and humidity, therefore they display robust potential when being utilized in thermal applications, like passive houses and low-energy greenhouses, as low auxiliary comfort will be required (Korjenic et al., 2016). Further benefits of utilizing natural fibers comprise good adhesion properties, renewability, low cost, availability, no pollution, or toxins are created, easy extraction, biodegradability, and high toughness. Besides, plant materials and fibers can retain dust, decrease noise levels, and can support as a shield against the sunlight and weathering effects when employed in the structural envelope. Plant fibers in composites can also help in reducing the normal dead load of a material because of lightweight. Cellulose, lignocellulose, and lignin are the terms used for natural fibers (Soares et al., 2016). With natural fibers, the environmental conditions at the time of harvest will impact their physical and mechanical properties. This comprises the soil condition, the extraction process, any treatment commenced, temperature, air humidity, and more. Natural fibers have numerous benefits. They are readily obtainable and are simply extracted, abundant, need little to no energy to produce, can be extracted from waste material, are economical and low-cost, and can lessen the environmental effect when utilized in the construction industry. Moreover, they have better thermal insulation properties, renewability, adhesion, no toxins or pollution are formed, and they have high toughness, are biodegradable, and lightweight (Arshad et al., 2016).

Manufacturing of polymeric composites reinforced with natural fibers has been an increasing trend and could substitute synthetic fibers for some applications, thus promoting sustainable materials and development (Petková et al., 2017).

The key limitations and disadvantages of natural fibers are that they naturally have reduced durability and may degrade with time. In addition, hydrophilicity is another disadvantage of plant fibers. Studies have been conducted to improve the durability of the fibers, which comprise a coating, chemical treatment, and replacing or blocking the hydroxyl group of the organic natural fibers, but more research is essential to conclude whether the enhancement in durability is long-term or just temporary. It has been indicated that these approaches are quite expensive and could be harmful as leeching, and exposure to chemical, biological, thermal, and UV-related degradation have not to be explored in detail. As natural fibers are likely to have a smaller lifespan, they may be limited and purposed as temporary reinforcement. Several natural fibers and their properties are discussed in subsections (Faghih et al., 2017).

1.5.1.1. Bast Fibers

Fibers including jute, flax, hemp, and kenaf can be classified and defined as bast fibers. These fibers are generally extracted from the exterior side of various plants. These fibers possess high tensile strength as well as have the ability to be a good thermal insulator (Figure 1.8) (Bos et al., 2019).

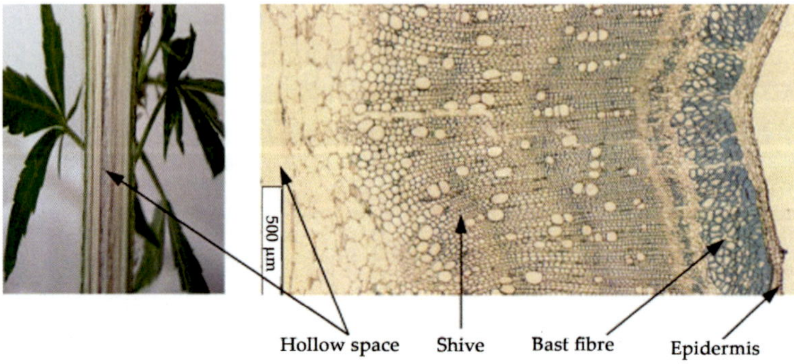

Figure 1.8. Cross-sectional view of bast fiber.

Source: https://www.sciencedirect.com/topics/engineering/bast-fiber.

Jute fibers are vegetable fibers and are one of the most studied natural fibers. Jute fibers are obtained from the exterior side of jute plants; the fibers are enclosed in the bark. Comparable to other materials, there are many types of jute fibers with different properties. These natural fibers have been utilized in pavement engineering (Mohajerani et al., 2017).

Flax fibers have been employed in composite materials to increase ductility and as reinforcement for building materials to enhance stabilization. Kenaf is a cellulose fiber that is readily obtainable in tropical and subtropical environments as in Malaysia (Arshad et al., 2016).

Korjernic et al. (2016) investigated the development of natural fibers-based insulation materials, principally for plant façade applications. Samples were prepared using a thermal binding technique, where the fibers of hemp, flax, and jute were used with the waste material called shive. Bi-component polyester fiber was utilized to adhere all the components together. The results revealed that the two samples with the preeminent results were the samples comprising 48% and 49% technical hemp fibers. The material that was developed had as good as thermal insulation properties to market and other natural fiber insulation products. The dynamic stiffness measured also displaced potential for applications like floorboards. It can be determined that if designed appropriately, the developed insulation material can be employed as external cladding for structures. Developments in terms of interior hydrothermal conditions, thermal comfort, and reduced humidity can be attained by employing this material (Laura et al., 2002; Suresh and Kumar, 2014).

1.5.1.2. Palm Tree Fibers

Palm tree fibers comprise oil palm fiber, coconut (coir) fibers, and date palm fiber. Palm fibers are low-cost, abundant, resilient, lightweight, high water absorption properties, and have comparative strength against deterioration. Palm fibers normally have low tensile strength and a low modulus of elasticity (Figure 1.9) (Deák and Czigány, 2009; Panda et al., 2017).

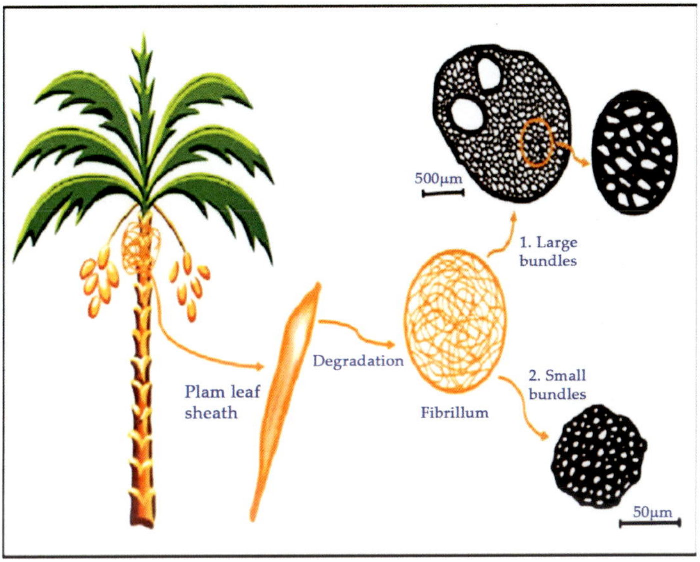

Figure 1.9. Schematic illustration of fiber extraction from date-palm tree.

Source: https://www.sciencedirect.com/science/article/abs/pii/S0141813019313510.

Coir fibers are obtained from the mesocarp section of the coconut. Coir fibers can be divided into two groups: short fibers if the length is below 130 mm and long fibers if the length surpasses 130 mm. These fibers are waste products for agricultural purposes that are persistently being created in the whole world. Coconut waste is commonly burnt or dumped but can be used productively. Coir fibers are strong, durable, tough, abundant, and cheap, having resistance against rot and fungi. Coir fibers are hard to burn, and have high elongation. They also possess optimum insulation properties, and, when wet, coir fibers sustain their tensile capacity. Various studies have investigated the physical and mechanical performance and behavior of coconut-reinforced soil (Dhand et al., 2015). Ghavami et al. (1999) revealed that adding 4% of coir fibers to soil blocks prohibited all evident cracks and amplified the ductility of the constructed product. As coir is comprised of high content of lignin, it has a prolonged service life compared to other natural fibers. This is because of the degradation process. As the fibers can be efficiently utilized for 4 to 10 years, they can be purposed as provisional reinforcement. It has been established that coir can support in improving the compressive strength, water absorption capability, and durability, and tensile strength of composite materials. These properties can also decline based on the volume content of coir used (Fiore et al., 2015).

Date palm fibers are water absorbent lightweight fibers that hold mechanical properties, like durability and tensile strength, and are cost-effective, easily available, and endure well against deterioration. These fibers can enhance the compressive as well as general strength of composite materials by hindering the movement of individuals and collections of particles. As stated before, the environmental condition and the time of harvest significantly affect the mechanical and physical performance of natural fibers. While date palm fibers are collected from a decayed parent tree, the fiber can be brittle, have low tensile strength, reduced modulus of elasticity, and high-water absorption property (Wang et al., 2008).

Scientists investigated the hydrophilicity and effect of moisture on the thermal conduction and density of a board manufactured from a mesh prepared from date palm fibers and mortar. The outcomes verified numerous benefits from utilizing date palm fiber mesh in the mortar. The date palm fiber improved the whole porosity, minimized the density and thermal conductivity, and, overall, produced a lighter material (Putman and Amirkhanian, 2004).

1.5.1.3. Cereal Straw

Straw is general material. The group 'cereal straw' denotes barley straw, wheat straw, oat straw, and straw. All over the world, straw is grown and harvested as a portion of agriculture. As revealed earlier, straw possesses a long past, as it was utilized as reinforcement to toughen mud bricks in the course of the Egyptian eras. Researchers have validated that if an optimal quantity of straw is utilized, positive effects will be shown by the reinforced material. Cereal straw has displayed the potential to decline shrinkage, lowering the curing time, increase ductility, and improve tensile and compressive strength. Straw has elevated water absorption properties (Pollack, 1972; Riggs, 1985). Researchers investigated the potential of barley and wheat straw for the structural industry. Certain promising benefits specified were its energy proficiency and enhanced resistance against fire. In this study, SEM technology was utilized to investigate ground samples of the straw. The results displayed that straw possesses low bulk density, which can support to the creation of lightweight materials and can considerably increase the hygrothermal feature of a structure (Figure 1.10) (Huang et al., 2007).

Figure 1.10. Different crockery items manufactured from wheat straw fibers.

Source: https://www.pinterest.com/pin/551409548109385037/.

1.5.1.4. Leaf Fibers

The group 'leaf fibers' represents sisal fibers, pineapple fibers, and banana fibers, and all three fibers are obtained from tropical plants. Sisal fibers are vegetable fibers, and, identical to jute fibers, are one of the extensively studied natural fibers. More than 57 species of sisal exist, where the fibers are acquired from the leaf part of the plant. The leaf can have a width of 6 to 10 cm and a length of 50 to 250 cm. The condition of the environment and soil, time of harvest, and type of sisal will affect the size of the leaves (Sakka, 1985; Butt et al., 2016).

Sisal fibers can be categorized into three classifications depending upon their length. If the length of the fiber is below 600 mm, it is called short fiber. If the length of the fiber is greater than 600 mm but is not more than

700 mm, it is termed a medium-sized fiber. Lastly, if the length of the fiber goes above 700 mm, it is a long fiber. The diameter of the fiber is from 0.06 mm to 0.4 mm. Sisal fibers have elevated water absorption capability of about 60% to 70%, and they were conventionally utilized to reinforce plaster sheets prepared from gypsum (Mohajerani et al., 2016).

Together with the coir fibers, Ghavami et al. revealed that the addition of 4% of sisal fibers to soil blocks eradicated all visible cracks, amplified the ductility of the construction product, and, in some measure, enhanced the compressive strength. Prabakar and Siridihar recognized that sisal fibers lessen the dry density of soil. Moreover, the shear stress and shear strength of the soil improved to some extent concurrently with an increase in the fiber concentration and length. Though, these properties start to decline when the limit was surpassed (Singer, 1979; Simon and Bunsell, 1984).

1.5.2. Synthetic Fibers

Synthetic fibers are artificially produced for a purpose. These comprise steel fibers, glass fibers, plastic fibers, both macro plastic fibers and microplastic fibers, and carbon fibers. Different fabrication techniques and base material composition will initiate different mechanical properties of the fibers (Parkyn, 1970; Kadir and Mohajerani, 2011).

Synthetic fibers can be classified into two types depending upon their geometry, these are micro synthetic fibers and macro synthetic fibers. The fibers are categorized based on Table 1.1 (Shindo, 1961; Smith et al., 1979).

Table 1.1. Differentiating Micro Synthetic Fibers with Macro Synthetic Fibers

SL. No.	Geometry	Micro Synthetic Fiber	Macro Synthetic Fiber
1.	Length	5–30 mm	30–60 mm
2.	Diameter/cross-section	5–100 μm	0.6–1 mm^2

Source: Yin et al. (2015).

Synthetic fibers are generally utilized as reinforcement to decrease plastic shrinkage and regulate flaws, substituting the customary wired mesh and steel bar reinforcement. By replacing the traditional reinforcement with synthetic fibers, the maintenance, labor, and construction cost are reduced, and time is saved from the construction workings. With the uninterrupted development of technology, and as more knowledge about synthetic fibers is deliberated, the types of synthetic fibers and their properties have frequently upgraded. Aimed at synthetic fibers, various production procedures and base material composition will bring about different mechanical properties of the fibers (Pollack, 1972; Riggs, 1985).

One benefit of synthetic fibers is that they can lessen the cost of construction as they can be utilized to replace traditional reinforcement such as steel mesh and steel rebars, which are considerably heavier and entail more energy, resources, and time to produce (Petrovic et al., 1985; Alani and Beckett, 2013).

On the other hand, among the drawbacks of synthetic fibers are that they are artificial, and therefore, need raw materials and energy. Depending on the production technique, discharges, and contamination will also be created, and in utmost circumstances, they are non-biodegradable and are problematic to recycle. As a result, indicating a negative influence on the environment. Different synthetic fibers and their properties are discussed in subsections (Perret and Ruland, 1970; Khan and Ali, 2016).

1.5.2.1. Polymeric Fibers

There are diverse types of polymeric fibers, comprising polypropylene (PP), nylon, and PE, which are extensively utilized in various applications. Polymeric fibers can be micro-plastic fibers or macro-plastic fibers. There are three foremost types of macro synthetic fibers: PP fibers, polyethylene terephthalate fibers (PET), and high-density polyethylene fibers (HDPE). The variety of fiber, and the mechanical and physical properties it retains, is interrelated to the method used to fabricate the material, as well as the geometry and aspect ratio of the raw materials, and even the properties of the recycled material if the fiber was prepared from a recycled product. They can be utilized in comparable applications to steel fibers, but they possess the advantage of being lighter in weight and resistant against corrosion (Northolt, 1981; Jiao et al., 2016).

Some procedures of manufacturing comprise the film sheet technique, melt spinning techniques, and manual cutting. The melt spinning technique can produce PP fibers and PE fibers having tensile strength more than 450 MPa. The film sheet technique can produce HDPE, PP, and PE terephthalate fibers. Recycled PET fibers have been manufactured using this technique having tensile strength and modulus of 420 MPa and 10 GPa, respectively. The manual cutting method is low-cost and can be utilized to manufacture recycled PET fibers; though, low-grade properties and fibers are created by this technique. Recycled PET fibers having a tensile strength of 160 MPa and a modulus of 3 GPa have been manufactured through the manual cutting method (Murday et al., 2017; Mohajerani et al., 2017).

Macro plastic fibers have gained enormous attention because of their prospective sustainability while compared to the extra energy-consuming steel reinforcement. PP fibers are amongst the extensively used and valued fibers and have benefits, like noncorrosive and resistance to chemicals, alkalis, and chlorides, making them an appropriate reinforcement fiber for a varied range of applications. Small quantity is required, making them cost-effective. PP fibers have pronounced mechanical properties comprising high tensile strength and high modulus of elasticity. A drawback of PP fibers is that they possess low density, which origins floating problems in some composite matrices. They also have low hydrophilic characteristics because of their hydrophobic nature, which could influence their capability to bond with certain matrices (Xue et al., 2013).

While compared to PP fibers, HDPE fibers possess slightly high density. In addition, they also hold a greater hydrophilic tolerance. Nevertheless, restrictions for applications arise, as PE has reduced tensile strength (Serin et al., 2012; Kumar et al., 2016).

Polyethylene terephthalate (PET) fibers possess the highest density among the three. They have better mechanical properties, like tensile strength and elastic modulus, and can be mixed easily in matrices such as concrete. A drawback of PET fibers is that more energy is required to manufacture them because of the production method; therefore, these fibers are more expensive. An additional drawback of PET fibers is that they are degradable in alkali environments, which restricts their range of applications. PET fibers can be manufactured using recycled bottles, which displays some sustainability potential. Recycled bottles are prepared from PET strings, which can be utilized to make PET fibers (Chen and Xu, 2010).

Nylon fibers are microplastic fibers that can increase the properties of the composite, like toughness, tensile strength, control shrinkage, and more. Though, microfibers do not exhibit as much effect as macro plastic fibers (Mirzababaei et al., 2018).

1.5.2.2. Steel Fibers

Like traditional steel reinforcement, the fundamental characteristic of steel fibers is their extraordinary tensile strength. Steel fibers have been frequently considered in concrete applications; therefore, they are generally utilized to enhance the mechanical properties of concrete (Figure 1.11) (Soltani et al., 2017).

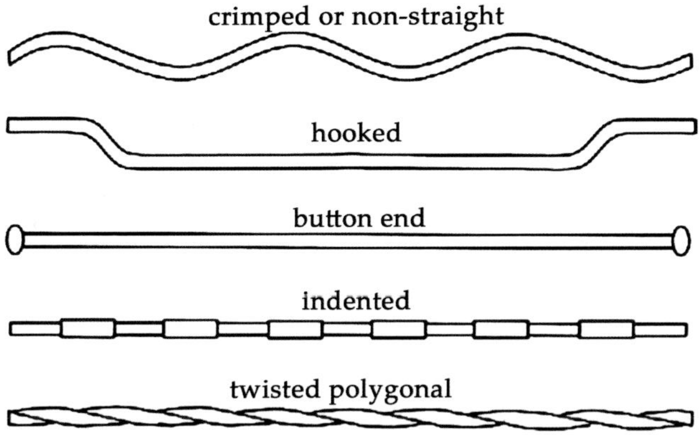

Figure 1.11. Different types of steel fibers.

Source: https://www.researchgate.net/figure/Types-of-Steel-Fibers_fig1_304734537.

Research has revealed that when steel fibers are utilized to reinforce concrete structures, there are numerous developments in the overall properties. Steel fibers assist to improve the concrete response in terms of ductility, toughness, shrinkage, cracking, resistance to fatigue, and impact and blast loading. Moreover, strength properties, like compressive strength, tensile strength, and flexural strength, are amplified for the parent material. This strength growth is because of the steel fibers' characteristics of absorbing energy and controlling cracks. Steel fibers can be an ultimate reinforcement to applications as they have better electric, magnetic, and heat conductivity (Mirzababae et al., 2016, 2019).

Though categorized with the common title of steel fibers, there are various kinds with different properties, and therefore, they are not all identical. Crimped fibers and hooked end are diverse kinds of steel fibers. Physical properties, like the length of fiber and fiber count, should also be considered while utilizing for particular applications. Moreover, the influence of adding steel fibers on the strength of the cast and printed concrete has been investigated. The research has revealed the fibers result in enhancement in flexural strength and eradicate the strength variance between cast and printed concrete without fibers (Morgan, 1979; Saad et al., 2012). Steel fibers can decrease the expense of construction as they can be utilized to replace the traditional reinforcement, like steel mesh and steel rebars, which are considerably heavier, and need more energy, resources, and time to manufacture. Funds are also preserved because of lower maintenance and labor required, and, additionally, the safety situations of the construction place will be enhanced. The drawback of steel fibers is analogous to steel reinforcement, in that it is fairly susceptible to corrosion, which may produce deficiency as well as deterioration of the parent material (Mirzababaei et al., 2012, 2018).

1.5.2.3. Glass Fibers

Numerous kinds of glass exist, with different chemical compositions, colors, and characteristics. Glass fibers have pronounced mechanical properties and surpass in terms of strength, durability, thermal properties, and have better interfacial bonding to the matrix. Glass fibers are most commonly utilized as reinforcement in resins and composites as they have incredible properties in strengthening composites (Binici et al., 2005; Mekhermeche et al., 2016).

Glass fibers are commonly utilized to reinforce PP systems. A composite is manufactured between the elements to create an exceptional material. The resultant composite is inexpensive, easily available, and retains the strength and toughness characteristics of glass fiber (Raut and Gomez, 2017).

Glass fibers can be used as reinforcement in polymeric materials. The resultant glass-fiber-reinforced polymeric material will have enhanced properties, like strength and stiffness, and the composite will also execute better performance in elevated temperature environments (Namboonruang and Yongam-Nuai, 2016).

In glass fiber reinforced composite materials, the mechanical performance of the composite materials strongly depends on the interfacial properties of the fiber-matrix interface. The premium composite properties are usually the consequence of a durable interface between the fiber and the matrix. Coupling agents that contribute to producing a durable bonded interface have been deliberated, and silane is a coupling agent that is frequently utilized for this reason. It assists bond and combines the glass fibers to the matrix, therefore producing a composite with the preeminent properties. Moreover, glass fiber-reinforced fly ash-based geopolymer concrete has been established for large-scale 3D printing uses with upgraded mechanical properties upon fiber mixing of up to 1% (Botev et al., 1999; Dash et al., 2010).

Glass fibers are manufactured through the spinneret technique, which is identical to the manufacturing of continuous basalt fibers. Typically, overhead gas burners are utilized to heat the melt during the production of glass fibers.

The melting point of the glass is about 1400–1600°C. Concerning basalt fibers, glass fibers are less impervious to alkali but perform better than basalt while encountered with acids (Figure 1.12) (Holbery and Houston, 2006; Fuentes et al., 2013).

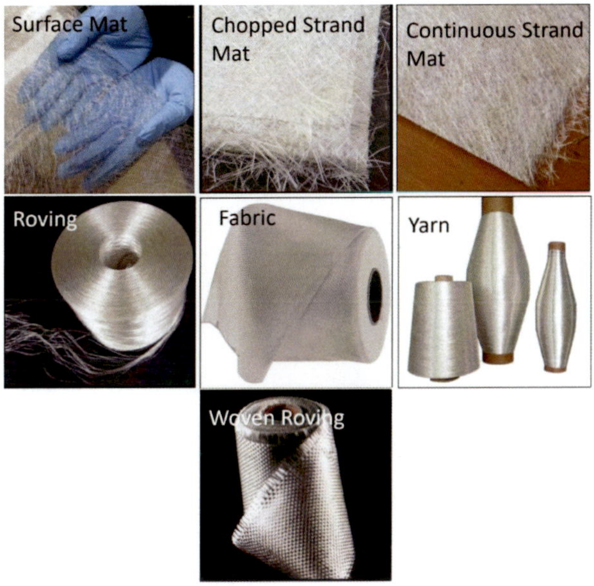

Figure 1.12. Different types of glass fibers.

Source: https://extrudesign.com/different-types-of-glass-fiber/.

E-glass is a common kind of glass fiber. E-glass is comprised of four major compounds:

- Al_2O_3 (aluminum oxide);
- SiO_2 (silicon dioxide);
- CaO (calcium oxide); and
- B_2O_3 (boric oxide).

Through earlier research studies, it has been believed that the Al_2O_3 (aluminum oxide) content significantly influenced the mechanical properties and behavior of the manufactured glass fiber. Because of this recognized information, there has been research increasing the overall Al_2O_3 (aluminum oxide) content (Milcwski et al., 1974, 1985).

Glass fiber composites have exceptional properties when deliberated and designed cautiously. Even though there are numerous benefits of utilizing glass fibers, some drawbacks are there as the production of glass fibers initiates concerns for the environment and uncertain sustainability. Manufacturing entails quantities of B_2O_3 (boric acid), which could not be sustainable as the compound is inadequate. The problem of the removal of glass products and glass fiber composites through the termination of life stage is also well known. Recycling could be a possibility for glass fiber composites, but most products are just burnt or concealed, which is not appropriate for the environment. An additional drawback of glass is that it degrades and has reduced resistance in alkali environments as it possesses poor alkali resistance. This restricts them from utilization in concrete applications, as concrete partakes an alkaline environment (Andersson and Warren, 1984).

1.5.2.4. Carbon Fibers

Carbon fibers have been used as reinforcement to produce composites with enhanced properties. The incorporation of carbon fibers produces a composite that has exceptional mechanical properties, performs well in elevated temperature environments, and has the advantage of durability. Though carbon fibers are relatively brittle, with vigilant deliberation in the design phase, carbon fiber-reinforced composites have exceptional properties.

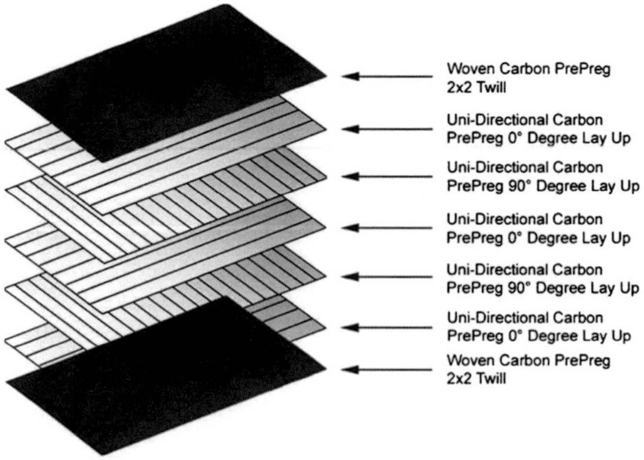

Figure 1.13. Different types of carbon fibers.

Source: https://www.jinjiuyi.net/news/what-is-carbon-fiber-sheet.html.

The drawbacks of carbon fibers are that because of their outstanding properties, the expenditure of manufacturing carbon fibers is high, and the strong bonding between the fibers and matrix material may be challenging to achieve (Figure 1.13) (Bennett et al., 1983; Bacon and Moses, 1986).

Like glass fibers, even if there are numerous positives and benefits to carbon fibers, the production of carbon fibers results in concerns for the environment and debatable sustainability. The difficulty to dispose of carbon fiber composites at the termination of life phase is also renowned. For major carbon fiber composites, recycling could be a prospect, yet maximum products are just burnt or buried, which is not good for the environment (Bennett and Johnson, 1979).

1.5.2.5. Aramid Fibers

Aramid fiber is a broad term for a class of synthetic organic fibers named aromatic polyamide (PA) fibers. The US Federal Trade Commission (FTC) provides a good explanation of an aramid fiber as "a manufactured fiber in which the fiber-forming substance is a long-chain synthetic PA in which at least 85% of the amide linkages are attached directly to two aromatic rings." Distinguished commercial names of aramid fibers comprise Kevlar and Nomex (Du Pont) and Twaron (Teijin Aramid). Teijinconex and Technora are two additional commercially available fibers from Teijin (Figure 1.14) (Magat, 1980; Mah et al., 1984).

Figure 1.14. Applications of aramid fibers.

Source: https://www.ipitaka.com/blogs/news/what-are-aramid-fibers-i.

Nylon is a general name for any long-chain PA. Aramid fibers such as Nomex or Kevlar, though, are ring compounds founded on the structure of benzene as opposed to the linear compounds employed to create nylon. The basic dissimilarity between Kevlar and Nomex is that Kevlar has para-oriented aromatic rings, i.e., its elementary unit is a symmetric molecule, with bonds from each aromatic ring being parallel, while Nomex is meta-oriented, with bonds at 120° angles to each other. Teijinconex is identical to Nomex while Twaron is similar to Kevlar. Technora is a copolyamide. The elementary chemical structure of aramid fibers comprises oriented para-substituted aromatic units, which makes them rigid rodlike polymers. The rigid rodlike structure gives rise to a high glass transition temperature and poor solubility, which makes manufacturing of these polymers, by conventional drawing procedures, challenging. Instead, they are spun from liquid crystalline polymer solutions by dry jet-wet spinning (Baker, 1983; Biro et al., 1992).

1.5.2.6. Ceramic Fibers

Continuous ceramic fibers show a smart package of properties. They combine high strength and elastic modulus with high-temperature proficiency and general liberty from environmental attack. These features make them significant reinforcements in high-temperature structural materials (Figure 1.15) (Lowrie, 1967).

Three methods are employed to manufacture ceramic fibers: chemical vapor deposition, polymer pyrolysis, and sol-gel techniques. The last two implicate rather novel techniques of attaining ceramics from organometallic polymers. The sol-gel technique is also utilized to manufacture a range of alumina-based oxide fibers commercially. An additional innovation in the ceramic fiber area is the idea of pyrolyzing; under organized conditions, polymers containing silicon and carbon or nitrogen can be transformed into high-temperature ceramic fibers. This idea is nothing but an extension of the polymer pyrolysis method to create multiple types of carbon fibers where an appropriate carbon-based polymer (e.g., PAN or pitch) is exposed to control heating to produce carbon fibers. The pyrolysis method of producing ceramic fibers has been utilized with polymers having nitrogen, silicon, carbon, and boron, with the end products being SiC, Si_3N_4, B_4C, and BN in fiber form, foam, or coating. We define some significant ceramic fibers (CJ and Scherer, 1990; Brown et al., 1992).

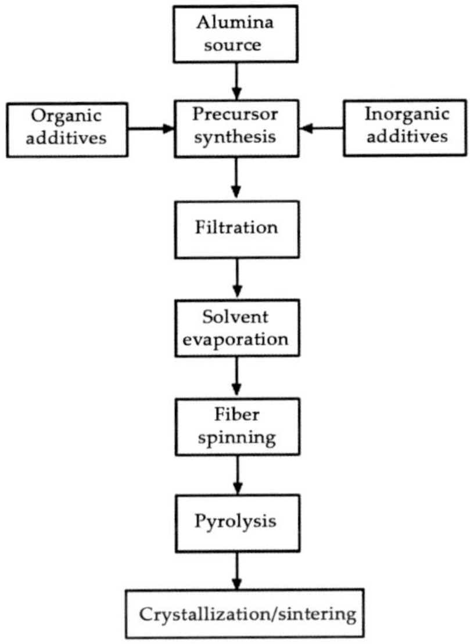

Figure 1.15. *Typical processing steps involved in the synthesis of a ceramic fiber.*

Source: https://www.nap.edu/read/6042/chapter/6.

1.5.2.7. Non-Oxide Fibers

Non-oxides like silicon carbide and silicon nitride have very substantial properties. Silicon carbide, particularly, commercially exists in a fibrous form (Loewenstein, 1973).

1.5.3. Basalt Fibers

Basalt is a fine-grained, dark, volcanic, and igneous mineral that is naturally found and easily obtainable on a global scale. It is a frozen rock made from the hot lava of volcanoes. In recent times, basalt has been extensively deliberated because of its strong potential as an appropriate fiber reinforcement for composite materials. Throughout several fields of study, sustainable materials have been comprehensively considered. This comprises research on the natural mineral of basalt, owing to its attractive and robust properties and low cost (Figure 1.16) (Lipowitz et al., 1990; Bunsell and Piant, 2006).

Figure 1.16. Different types of basalt fibers.

Source: http://newwaytechnologies.nl/2019/10/01/basalt-fibers/.

Basalt has a fairly extended history. Since Roman times, basalt rock has been utilized in its normal state for concrete surfaces and as stones used in building. Scientists in the United States investigated the rock in the early 1900s when substantial research was done. During the mid-1900s, the United States and the Ukraine Soviet Union comprehensively examined the properties of basalt fibers, and, explicitly, their prospective use as a composite material in defense and military applications. Having been extensively explored in the civil engineering industry, interest has the extent to various ranges of engineering fields. With an extended history, the research of basalt fibers is yet a trend nowadays (Ciferri and Ward, 1979; Li et al., 1992).

Comprising of a range of oxides and alumina, basalt is a natural mineral-rich in chemicals. The central compounds of basalt in order of volume are SiO_2 (silicon dioxide), Al_2O_3 (aluminum oxide), Fe_2O_3 (ferric oxide), FeO (ferrous oxide), CaO (calcium oxide), and, lastly, MgO (magnesium oxide). The chemical composition of the basalt rock affects the thermal, chemical, and mechanical properties it holds containing the strength, physical appearance, density, heat conduction, and thermal stability. Based on chemical composition, basalt is categorized according to Table 1.2 (Chawla, 2012).

Table 1.2. Classification of Basalt

SL. No.	Classification	Content of Silicon Dioxide
1.	Acidic basalt	$SiO_2 \geq 46\%$
2.	Mildly acidic basalt	$42\% \leq SiO_2 < 46\%$
3.	Alkaline basalt	$SiO_2 < 42\%$

Source: Soares et al. (2016).

Basalt fibers can be distributed into two classifications: short basalt fiber and continuous basalt fiber. These two kinds of basalt fibers are manufactured through different approaches and hold different degrees of properties. The foremost differences between the fibers can be witnessed in Table 1.3 (Chawla, 1981, 2012).

Table 1.3. Difference between Short and Continuous Basalt Fiber

SL. No.	Differences	Continuous Basalt Fiber	Short Basalt Fiber
1.	Production method	Spinneret method	Melt blowing/ Junker's method
2.	Deviances in properties	Uniform	Uneven
3.	Properties	Stronger	Weaker
4.	Quality	Higher	Lower
5.	Cost	Higher	Lower

Source: Hafsa and Rajesh (2015); and Deák and Czigány (2009).

Basalt fibers display a significant potential because of their exceptional chemical, physical, and mechanical properties, which make them competitive candidates as reinforcement for composites. Some of their strong properties are revealed below (De Lamotte and Perry, 1970; Chawla and Bastos, 1979):

- Optimum electric, electromagnetic, and sound insulation properties;
- Better resistance and stability against impact load, corrosion, chemical attack, and fire;

- Good adhesion and abrasion properties with the ability to combine well with matrix materials;
- Extraordinary tensile strength and modulus of elasticity;
- High resistance to weather and acidic environments, and some alkali resistance;
- Absorb sound and vibration isolation;
- Resistant to radiation and UV light;
- Strong, hard, and rigid;
- Improved strain failure;
- Good thermal properties and stability, and can endure and perform well in elevated temperature environments;
- Nonreactive and noncombustible;
- Low moisture/water absorption and thermal conductivity.
- Basalt fibers are also:
- Easy to produce and process;
- Cost-efficient/inexpensive;
- Natural;
- Biodegradable;
- Ecologically clean, easy to handle, and non-toxic;
- Can be used in a different range of applications;
- Titled as green;
- Used to form lightweight composites with excellent properties;
- Can be recycled for reuse;
- Entail no chemicals or additives.

As stated in Table 1.3, short fibers and continuous fibers are prepared using different procedures and possess different properties. Short fibers are manufactured using the melt blowing technique and continuous fibers are made using the spinneret method. The basalt rock must be categorized as acidic basalt with over 46% of silicon dioxide to be appropriate to manufacture continuous fibers (DeBolt, 1982; DeTeresa et al., 1984).

Short basalt fibers can be manufactured through a single step using crushed basalt rock by melt blowing technique. In this method, the crushed basalt is melted and transferred on top of a set of centrifugal steel cylinders. Through the air pressure created by air jets, the melt is blown off from the cylinders and solidifies into fibers. Through the spinneret technique, improved quality continuous fibers are created, but this method involves more time and energy. Crushed basalt stone is melted in a rhodium-platinum pot and the melt is then streamed to a spinneret that is made from an equal material. The spinneret process will then activate, which will produce a spool of continuous basalt fiber. It can be noticed that no compound was added to the molten basalt in either method of production. Through diverse methods, short, and continuous fibers with different properties can be manufactured (Diefendorf and Tokarsky, 1975; DiCarlo, 1985).

1.6. WASTE MATERIALS AND FIBERS

The upsurge in waste material volume with inadequate landfill space is a renowned issue that continues to intensify due to the growth in the population. By using waste materials and fibers, the volume of waste that terminates in the landfill can be significantly reduced, and the energy used for the combustion process can also be preserved. In this section, the kinds and potential of waste materials and fibers are deliberated (Dobb et al., 1980).

Waste materials comprise fibers that could be consumed in applications as a lucrative and sustainable choice. Waste materials are categorized as secondary materials, as they are recycled, however, they display potential in substituting primary virgin fibers for particular applications. Raw energy, resources, and the extraction processes can be saved that is essential while acquiring key materials and fibers. This will also result in a decline in pollution and emissions (Dresher, 1969).

Tires and carpets are items that can be used for their fibers. In the case of tires, the rubber material and steel constituents can be reprocessed, which leaves the polyester fiber residual. The rubber element of the tire is light in weight, possesses low thermal conductivity, extraordinary hydraulic conductivity, and has an optimum shear strength when the strain is high. Unwanted carpets can be cut into small pieces and utilized as polyester fibers. These fibers can assist to enhance the bulk performance of the composite material by improving properties like strength and durability (Lee and JG, 1975; Svedberg et al., 1998).

Natural hair can be categorized as both a natural and a waste product. The fiber of hair can be utilized as reinforcement in composites to enhance the properties like strength, stability, and bearing capability, in addition, to reduce the development of flaws. Further natural fibers comprising coconut fibers are also a surplus material (Ezekiel and Spain, 1967).

Glass is a material that can be persistently reprocessed and reused while preserving extraordinary characteristics and properties. Though, there are certain restrictions to the capacity that can be recycled. For instance, in the United Kingdom, merely a third of residual glass materials are reprocessed because of limits on the infrastructure utilized for the collection process. If the collection process is improved around the world, more waste glass can be recycled. The recycling of waste glass materials for glass-making applications assists to preserve the raw materials, energy, and resources that are usually consumed for primary glass. There are numerous kinds of glass with different chemical compositions, colors, and characteristics, and the color of the waste glass determines whether it is suitable for recycling and re-use. Glass material that cannot be recycled because of its color can strive for different industries, like the construction and civil industry (Flory, 1956).

Steel slag is a by-product obtained from the steel industry. Unlike waste glass and other waste materials, steel slag can be simply acquired from the controlled environment of a steel development plant. Steel slag partakes better mechanical properties, irregular surface texture, hardness, and has an angular shape. The specific gravity of steel slag is approximately 3.2–3.6 g/cm^3 (Laine and Babonneau, 1993).

Investigation and analysis of recycled plastic fibers have been commenced, but there exists a risk if the preceding history of the plastic is not identified. If it has been acquired from an uncontrolled environment, the resultant properties of the recycled fiber may be unbalanced and unpredictable. Recycled plastics have been utilized to manufacture products comprising insulation products, ducts, pipes, furniture, and more. Waste recycled plastic bottles are also used to produce PE terephthalate fibers (Galasso and Paton, 1966; Fourdeux et al., 1971).

A different waste material that is in abundance and cannot be easily disposed of is cigarette butts. Cigarette butts display potential in producing unadorned destruction to the environment, as cigarette smoke comprises over 60 cancerous chemicals, where some may be trapped in the butts. These captured chemicals could probably infect the surrounding area, as the chemicals may leech out. This makes cigarette butts challenging to recycle. The filter of a cigarette comprises cellulose acetate. Cigarette fibers can be deliberated to be cellulose fibers, as they have been produced from extremely refined cellulose fibers that are attained from wood pulp, by acetylation. The discarding of cigarette butts in landfills and the burning of the waste product are not workable, economically proficient, or harmless. Therefore, studies on the integration and uses of cigarette butts recycled in particular applications have been deliberated by different researchers (Galasso et al., 1967; Gasson and Cockayne, 1970).

Waste materials comprise fibers that could be used in applications as an inexpensive, eco-friendly, and sustainable opportunity. There are numerous benefits of using waste materials and fibers in construction materials. The amount of waste that ends up in landfills can be significantly reduced; the energy, resources, and extraction processes that are mandatory when acquiring virgin materials can be well-preserved; and the energy utilized for the incineration of landfill materials can also be conserved. There will also be a decrease in pollution and emissions, which is good for the environment (Haggerty, 1972; Hild and Schwartz, 1992a, b).

Different waste materials, such as tires, glass, steel slag, carpet, natural hair, coconut fiber, plastic, and cigarette butts, were discoursed in this review. Through numerous research studies, it has been explored that there are benefits in using waste materials in construction materials as a sustainable and inexpensive solution for certain applications (Laffon et al., 1989).

In SMA mixtures, waste tire and carpet materials improved the durability and tensile strength, stopped down the drain from happening, and delivered steadiness to the mixture. It offers a cost outcome substitute and exhibits sustainable value. Preceding experiments and investigations have asserted that while recycling crumb rubber in the concrete slab, the slab enhanced the fire resistance performance, and the spalling deficiencies produced by the fire were reduced. Waste carpet fibers have been utilized as reinforcement in clay soils as they increase the shear strength, the stress distributor, operative stress ratio, the effective cohesion, the free-range compression strength (UCS), and compact swelling. Waste carpet fibers have also been effectively utilized to increase the bearing resistance properties of slopes (Hodd et al., 1978; Inal et al., 1980).

Natural hair was examined, and it enhanced the strength and stability of the clay soil. The prime fiber content was 2%. It has been recommended that hair could probably be utilized to reinforce embankments and support stabilize slopes. Coconut fibers have been utilized to reinforce and rise the strength of soil, where the optimum fiber content was 0.75%. Steel slag displayed the potential to be a pronounced replacement for rough aggregates in surface asphalt.

In conclusion, it has been determined that cigarette butts could productively be incorporated in fired clay bricks and asphalt concrete when encapsulated. These research studies indicate how to discard and effectively use toxic waste materials (Jaffe et al., 1985; Johnson and Tyson, 1969).

The drawbacks of utilizing waste materials are determined by the application in which they are being utilized. In asphalt pavements, there are apprehensions over the prospective effects, such as excess pollutants and leaching. Apprehensions also rise while utilizing industrial waste, like steel slag, due to such properties as electrical conductivity and concentration of chromium. The processing cost for reutilizing the waste materials may frequently be greater than the characteristic cost of virgin materials, which confines their application for usage (Kaplan et al., 1988; Kumashiro et al., 2010).

1.7. IMPORTANT FACTORS TO CONSIDER WHEN USING FIBERS

Through different research and investigations, it has been revealed that the quantity of fibers utilized to reinforce the composite material is crucial. The higher proportion of fiber is not essentially better. Though fiber reinforcement is eminent for improving the mechanical properties of materials, this is typically only up to a certain amount, and, if surpassed, the properties will show degradation. This is why the optimum fiber content within a composite material is significant for obtaining better results. Physical properties, like the fiber count, the fiber length, and the distribution of fiber, should be considered. In concrete, by increasing the quantity of fibers, the workability of the concrete mixture declines, therefore additives may be required. Moreover, in soil short fibers were described to be more operative. Treatment may also be done to enhance the properties of fibers like alkaline treatment, water repellence, and surface development (Krukonis, 1977; Kwolek et al., 1977).

Fibers have been utilized to manufacture high-performance composites with further matrices. The physicochemical interaction between the constituents of the composite, their adhesion, and the bond between the fibers and the matrix are exceptionally key factors that impact the mechanical performance of the composite. The optimum composite properties are characteristically the consequence of a strong interfacial bond between the fibers and the matrix. Additives, like coupling agents, have been employed to increase the adhesion between the composites to generate durable bonds. They assist interfacial bond and unite the fibers with the matrix, hence creating a composite with the finest properties (LaBelle, 1971; Yang, 1993).

REFERENCES

1. Alani, A. M., & Beckett, D., (2013). Mechanical properties of a large-scale synthetic fiber reinforced concrete ground slab. *Constr. Build. Mater., 41*, 335–344. [CrossRef.].
2. Altoubat, S. A., Roesler, J. R., Lange, D. A., & Rieder, K. A., (2008). Simplified method for concrete pavement design with discrete structural fibers. *Constr. Build. Mater., 22*, 384–393. [CrossRef.].
3. Andersson, C. H., & Warren, R., (1984). Silicon carbide fibers and their potential for use in composite materials; Part 1. *Composites, 15*(1), 16–24.
4. Arshad, A. K., Mansor, S., Shafie, E., & Hashim, W., (2016). Performance of stone mastic asphalt mix using selected fibers. *J. Teknol., 78*, 99–103. [CrossRef.].
5. Bacon, R., & Moses, C. T., (1986). Carbon fibers, from light bulbs to outer space. In: *High Performance Polymers: Their Origin and Development* (Vol. 1, pp. 341–353). Springer, Dordrecht.
6. Baker, A. A., (1983). High-performance fiber composites for aircraft applications. An overview. In: *Metals Forum* (Vol. 6, No. 2, pp. 81–101).
7. Barham, L. P., & Keller, A., (1985). High-strength polyethylene fibers from solution and gel spinning. *Journal of Materials Science, 20*(7), 2281–2302.
8. Behfarnia, K., & Behravan, A., (2014). Application of high-performance polypropylene fibers in concrete lining of water tunnels. *Mater. Des., 55*, 274–279. [CrossRef.].
9. Bennett, S. C., & Johnson, D. J., (1979). Electron-microscope studies of structural heterogeneity in pan-based carbon fibers. *Carbon, 17*(1), 25–39.
10. Bennett, S. C., Johnson, D. J., & Johnson, W., (1983). Strength-structure relationships in PAN-based carbon fibers. *Journal of Materials Science, 18*(11), 3337–3347.
11. Binici, H., Aksogan, O., & Shah, T., (2005). Investigation of fiber-reinforced mud brick as a building material. *Constr. Build. Mater., 19*, 313–318. [CrossRef.].
12. Biro, D. A., Pleizier, G., & Deslandes, Y., (1992). Application of the microbond technique part III effects of plasma treatment on the

ultra-high modulus polyethylene fiber-expoxy interface. *Journal of Materials Science Letters, 11*(10), 698–701.

13. Bos, F. P., Bosco, E., & Salet, T. A., (2019). Ductility of 3D printed concrete reinforced with short straight steel fibers. *Virtual Phys. Prototyp., 14*, 160–174. [CrossRef.].
14. Botev, M., Betchev, H., Bikiaris, D., & Panayiotou, C., (1999). Mechanical properties and viscoelastic behavior of basalt fiber-reinforced polypropylene. *J. Appl. Polym. Sci., 74*, 523–531. [CrossRef.].
15. Bouasker, M., Belayachi, N., Hoxha, D., & Al-Mukhtar, M., (2014). Physical characterization of natural straw fibers as aggregates for construction materials applications. *Materials, 7*, 3034–3048. [CrossRef.] [PubMed.].
16. Boukhattem, L., Boumhaout, M., Hamdi, H., Benhamou, B., & Nouh, F. A., (2017). Moisture content influence on the thermal conductivity of insulating building materials made from date palm fibers mesh. *Constr. Build. Mater., 148*, 811–823. [CrossRef.].
17. Brown, J. R., Chappell, P. J. C., & Mathys, Z., (1992). Plasma surface modification of advanced organic fibers. *Journal of Materials Science, 27*(12), 3167–3172.
18. Bunsell, A. R., & Piant, A., (2006). A review of the development of three generations of small diameter silicon carbide fibers. *Journal of Materials Science, 41*(3), 823–839.
19. Butt, W. A., Mir, B. A., & Jha, J. N., (2016). Strength behavior of clayey soil reinforced with human hair as a natural fiber. *Geotech. Geol. Eng., 34*, 411–417. [CrossRef.].
20. Chawla, K. K., & Bastos, A. C., (1979). *Proc. 3rd. Int. Conf. Mech. Behavior of Materials* (Vol. 3, pp. 1–32).
21. Chawla, K. K., (1981). Composite materials 1980: A report on the 3rd international conference on composite materials, Paris. *Materials Science and Engineering, 48*(1), 137–141.
22. Chawla, K. K., (2012). *Composite Materials: Science and Engineering* (Vol. 1, pp. 1–33). Springer Science & Business Media.
23. Chen, H., & Xu, Q., (2010). Experimental study of fibers in stabilizing and reinforcing asphalt binder. *Fuel, 89*, 1616–1622. [CrossRef.].
24. Ciferri, A., & Ward, I. M., (1979). *Ultra-High Modulus Polymers* (Vol. 1, pp. 1–33). Applied Science Publishers.

25. CJ, B., & Scherer, G. W., (1990). *Sol-Gel Science: The Physics and Chemistry of Sol-Gel Processing*. New York, Academic.
26. Dash, A. K., Thatoi, D. N., & Sarangi, M. K., (2010). Analysis of the mechanical characteristics of a red mud filled hybridized composite. In: *Proceedings of the Frontiers in Automobile and Mechanical Engineering (FAME)* (Vol. 1, pp. 1–33). Chennai, India.
27. De Lamotte, E., & Perry, A. J., (1970). Diameter and strain-rate dependence of the ultimate tensile strength and young's modulus of carbon fibers. *Fiber Science and Technology, 3*(2), 157–166.
28. Deák, T., & Czigány, T., (2009). Chemical composition and mechanical properties of basalt and glass fibers: A comparison. *Text. Res. J., 79*, 645–651. [CrossRef.].
29. DeBolt, H. E., (1982). Boron and other high-strength, high-modulus, low-density filamentary reinforcing agents. In: *Handbook of Composites* (Vol. 1, pp. 171–195). Springer, Boston, MA.
30. DeTeresa, S. J., Allen, S. R., Farris, R. J., & Porter, R. S., (1984). Compressive and torsional behavior of Kevlar 49 fiber. *Journal of Materials Science, 19*(1), 57–72.
31. Dhand, V., Mittal, G., Rhee, K. Y., Park, S. J., & Hui, D., (2015). A short review on basalt fiber reinforced polymer composites. *Compos. Part B Eng., 73*, 166–180. [CrossRef.].
32. DiCarlo, J. A., (1985). Fibers for structurally reliable metal and ceramic composites. *JOM, 37*(6), 44–49.
33. Diefendorf, R. J., & Tokarsky, E., (1975). High-performance carbon fibers. *Polymer Engineering & Science, 15*(3), 150–159.
34. Dobb, M. G., Johnson, D. J., & Saville, B. P., (1980). Structural aspects of high modulus aromatic polyamide fibers. *Philosophical Transactions of the Royal Society of London: Series A, Mathematical and Physical Sciences, 294*(1411), 483–485.
35. Dresher, W. H., (1969). The age of fibers. *JOM, 21*(4), 17–26.
36. Ezekiel, H. M., & Spain, R. G., (1967). Preparation of graphite fibers from polymeric fibers. In: *Journal of Polymer Science Part C: Polymer Symposia* (Vol. 19, No. 1, pp. 249–265). New York: Wiley Subscription Services, Inc., A Wiley Company.
37. Faghih, F., Das, D., & Ayoub, A., (2017). Seismic behavior of fiber-reinforced steel-concrete composite systems. *Procedia Eng., 171*, 899–908. [CrossRef.].

38. Fiore, V., Scalici, T., Di Bella, G., & Valenza, A., (2015). A review on basalt fiber and its composites. *Compos. Part B Eng., 74*, 74–94. [CrossRef.].
39. Flory, P. J., (1956). Phase equilibria in solutions of rod-like particles. *Proceedings of the Royal Society of London: Series A. Mathematical and Physical Sciences, 234*(1196), 73–89.
40. Fourdeux, A., Perret, R., & Ruland, W., (1971). *Carbon Fibers: Their Composites and Applications* (Vol. 1, pp. 1–32). London: The Plastic Institute.
41. Fuentes, C., Tran, L., Van, H. M., Janssens, V., Dupont-Gillain, C., Van, V. A., & Verpoest, I., (2013). Effect of physical adhesion on mechanical behavior of bamboo fiber-reinforced thermoplastic composites. *Colloids Surf. A Physicochem. Eng. Asp., 418*, 7–15. [CrossRef.].
42. Galasso, F., & Paton, A., (1966). *Tungsten Borides in Boron Fibers* (Vol. 1, pp. 1–26). United Aircraft Corp., East Hartford, Conn.
43. Galasso, F., Kuehl, D., & Tice, W., (1967). Amorphous boron fibers. *Journal of Applied Physics, 38*(1), 414, 415.
44. Gasson, D. B., & Cockayne, B., (1970). Oxide crystal growth using gas lasers. *Journal of Materials Science, 5*(2), 100–104.
45. Ghavami, K., Filho, R. D. T., & Barbosa, N. P., (1999). Behavior of composite soil reinforced with natural fibers. *Cem. Concr. Compos., 21*, 39–48. [CrossRef.].
46. Haggerty, J. S., (1972). *Production of Fibers by a Floating Zone Fiber Drawing Technique, 1*, 1–32.
47. Hejazi, S. M., Sheikhzadeh, M., Abtahi, S. M., & Zadhoush, A., (2012). A simple review of soil reinforcement by using natural and synthetic fibers. *Constr. Build. Mater., 30*, 100–116. [CrossRef.].
48. Hild, D. N., & Schwartz, P., (1992a). Plasma-treated ultra-high-strength polyethylene fibers; Part I. Characterization by electron spectroscopy for chemical analysis. *Journal of Adhesion Science and Technology, 6*(8), 879–896.
49. Hild, D. N., & Schwartz, P., (1992b). Plasma-treated ultra-high-strength polyethylene fibers; Part II. Increased adhesion to poly(methyl methacrylate). *Journal of Adhesion Science and Technology, 6*(8), 897–917.
50. Hodd, K. A., KA, H., & DC, T., (1978). *High Modulus Fibers, 1*, 1–20.

51. Holbery, J., & Houston, D., (2006). Natural-fiber-reinforced polymer composites in automotive applications. *JOM, 58*, 80–86. [CrossRef.].
52. Huang, Y., Bird, R. N., & Heidrich, O., (2007). A review of the use of recycled solid waste materials in asphalt pavements. *Resour. Conserv. Recycl., 52*, 58–73. [CrossRef.].
53. Hurley, G. F., & Pollock, J. T. A., (1972). Mechanical and structural characterization of sapphire ribbons and continuous filaments. *Metallurgical and Materials Transactions B, 3*(2), 401–406.
54. Inal, O. T., Leca, N., & Keller, L., (1980). Characterization of polyacrylonitrile based carbon fibers. *Physica Status Solidi (a), 62*(2), 681–693.
55. Jaffe, M., Jones, R. S., Lewin, M., & Preston, S., (1985). *Handbook of Fiber Science and Technology* (Vol. 1, pp. 340–349). High technology fibers. Part A. Marcel Dekker, New York.
56. Jamshaid, H., & Mishra, R., (2016). A green material from rock: Basalt fiber: A review. *J. Text. Inst., 107*, 923–937. [CrossRef.].
57. Jiao, L., Su, M., Chen, L., Wang, Y., Zhu, H., & Dai, H., (2016). Natural cellulose nanofibers as sustainable enhancers in construction cement. *PLoS ONE, 11*, e0168422. [CrossRef.].
58. Johnson, D. J., & Tyson, C. N., (1969). The fine structure of graphitized fibers. *Journal of Physics D: Applied Physics, 2*(6), 787.
59. Jones, R. W., (1990). Fundamental principles of sol-gel technology. *The Institute of Metals, 1990*, 128.
60. Kadir, A. A., & Mohajerani, A., (2011). Recycling cigarette butts in lightweight fired clay bricks. *Proc. Inst. Civ. Eng. Constr. Mater., 164*, 219–229. [CrossRef.].
61. Kadir, A. A., Zulkifly, S. N. M., Abdullah, M. M. A. B., & Sarani, N. A., (2016). The utilization of coconut fiber into fired clay brick. *Key Eng. Mater., 673*, 213–222. [CrossRef.].
62. Kaplan, S. L., Rose, P. W., Nguyen, H. X., & Chang, H. W., (1988). Gas plasma treatment of spectra fiber. *Materials-Pathway to the Future, 1*, 551–559.
63. Khan, M., & Ali, M., (2016). Use of glass and nylon fibers in concrete for controlling early age micro cracking in bridge decks. *Constr. Build. Mater., 125*, 800–808. [CrossRef.].

64. Korjenic, A., Zach, J., & Hroudová, J., (2016). The use of insulating materials based on natural fibers in combination with plant facades in building constructions. *Energy Build., 116*, 45–58. [CrossRef.].

65. Krukonis, V., (1977). Chemical vapor deposition of boron filament. In: *Boron and Refractory Borides* (Vol. 1, pp. 517–540). Springer, Berlin, Heidelberg.

66. Kumar, T. M., Ramesh, A., & Kumar, M., (2016). Evaluation of rutting characteristics in stone mastic asphalt mix when added with basalt fiber. *I Manag. J. Mater. Sci., 4*, 19.

67. Kumashiro, Y., Yamato, M., & Okano, T., (2010). Cell attachment-detachment control on temperature-responsive thin surfaces for novel tissue engineering. *Annals of Biomedical Engineering, 38*(6), 1977–1988.

68. Kwolek, S. L., Morgan, P. W., Schaefgen, J. R., & Gulrich, L. W., (1977). Synthesis, anisotropic solutions, and fibers of poly(1,4-benzamide). *Macromolecules, 10*(6), 1390–1396.

69. Labelle, H. E., & Mlavsky, A. I., (1967). Growth of sapphire filaments from the melt. *Nature, 216*(5115), 574–575.

70. LaBelle, Jr. H. E., (1971). Growth of controlled profile crystals from the melt: Part II-edge-defined, film-fed growth (EFG). *Materials Research Bulletin, 6*(7), 581–589.

71. Laborel-Préneron, A., Aubert, J., Magniont, C., Tribout, C., & Bertron, A., (2016). Plant aggregates and fibers in earth construction materials: A review. *Constr. Build. Mater., 111*, 719–734. [CrossRef.].

72. Laffon, C., Flank, A. M., Lagarde, P., Laridjani, M., Hagege, R., Olry, P., & Legrand, A. P., (1989). Study of Nicalon-based ceramic fibers and powders by EXAFS spectrometry, X-ray diffractometry and some additional methods. *Journal of Materials Science, 24*(4), 1503–1512.

73. Laine, R. M., & Babonneau, F., (1993). Preceramic polymer routes to silicon carbide. *Chemistry of Materials, 5*(3), 260–279.

74. Laura, D., Keskkula, H., Barlow, J., & Paul, D., (2002). Effect of glass fiber surface chemistry on the mechanical properties of glass fiber reinforced, rubber-toughened nylon 6. *Polymer, 43*, 4673–4687. [CrossRef.].

75. Lee, J. G., & JG, L., (1975). *Formation of Silicon Carbide from Rice Hulls, 1*, 1–33.

76. Li, Z. F., Netravali, A. N., & Sachse, W., (1992). Ammonia plasma treatment of ultra-high-strength polyethylene fibers for improved adhesion to epoxy resin. *Journal of Materials Science, 27*(17), 4625–4632.
77. Lipowitz, J., Rabe, J. A., Frevel, L. K., & Miller, R. L., (1990). Characterization of nano porosity in polymer-derived ceramic fibers by x-ray scattering techniques. *Journal of Materials Science, 25*(4), 2118–2124.
78. Lloyd, S., (2014). Steel fibers in concrete floor slabs. *Concr. Int., 36*, 47–49.
79. Loewenstein, K. L., (1973). *The Manufacturing Technology of Continuous Glass Fibers, 1*, 1–27.
80. Lowrie, R. E., (1967). Glass fibers for high strength composites. *Modern Composite Materials, 1*, 270–321.
81. Magat, E. E., (1980). Fibers from extended chain aromatic polyamides. *Philosophical Transactions of the Royal Society of London. Series A, Mathematical and Physical Sciences, 294*(1411), 463–472.
82. Mah, T., Hecht, N. L., McCullum, D. E., Hoenigman, J. R., Kim, H. M., Katz, A. P., & Lipsitt, H. A., (1984). Thermal stability of SiC fibers (Nicalon®). *Journal of Materials Science, 19*(4), 1191–1201.
83. Mekhermeche, A., Kriker, A., & Dahmani, S., (2016). Contribution to the study of thermal properties of clay bricks reinforced by date palm fiber. In: *AIP Conference Proceedings* (Vol. 1, pp. 1–33). AIP Publishing: Melville, NY, USA.
84. Milcwski, J. V., Sandstrom, J. L., & Brown, W. S., (1974). *Silicon Carbide 1973, 1*, 1–20.
85. Milewski, J. V., Gac, F. D., Petrovic, J. J., & Skaggs, S. R., (1985). Growth of beta-silicon carbide whiskers by the VLS process. *Journal of Materials Science, 20*(4), 1160–1166.
86. Mirzababaei, M., Arulrajah, A., Haque, A., Nimbalkar, S., & Mohajerani, A., (2018). Effect of fiber reinforcement on shear strength and void ratio of soft clay. *Geosynth. Int., 25*, 471–480. [CrossRef.].
87. Mirzababaei, M., Miraftab, M., Mohamed, M., & McMahon, P., (2012). Unconfined compression strength of reinforced clays with carpet waste fibers. *J. Geotech. Geoenvironmental Eng., 139*, 483–493. [CrossRef.].

88. Mirzababaei, M., Miraftab, M., Mohamed, M., & McMahon, P., (2013). Impact of carpet waste fiber addition on swelling properties of compacted clays. *Geotech. Geol. Eng., 31*, 173–182. [CrossRef.].
89. Mirzababaei, M., Mohamed, M., & Miraftab, M., (2016). Analysis of strip footings on fiber-reinforced slopes with the aid of particle image velocimetry. *J. Mater. Civ. Eng., 29*, 04016243. [CrossRef.].
90. Mirzababaei, M., Mohamed, M., Arulrajah, A., Horpibulsuk, S., & Anggraini, V., (2018). Practical approach to predict the shear strength of fiber-reinforced clay. *Geosynth. Int., 25*, 50–66. [CrossRef.].
91. Mirzababaei, M., Yasrobi, S., & Al-Rawas, A., (2009). Effect of polymers on swelling potential of expansive soils. *Proc. Inst. Civ. Eng. Ground Improv., 162*, 111–119. [CrossRef.].
92. Mohajerani, A., Kadir, A. A., & Larobina, L., (2016). A practical proposal for solving the world's cigarette butt problem: Recycling in fired clay bricks. *Waste Manag., 52*, 228–244. [CrossRef.] [PubMed.].
93. Mohajerani, A., Tanriverdi, Y., Nguyen, B. T., Wong, K. K., Dissanayake, H. N., Johnson, L., Whitfield, D., et al., (2017). physico-mechanical properties of asphalt concrete incorporated with encapsulated cigarette butts. *Constr. Build. Mater., 153*, 69–80. [CrossRef.].
94. Mohajerani, A., Vajna, J., & Ellcock, R., (2018). Chromated copper arsenate timber: A review of products, leachate studies and recycling. *J. Clean. Prod., 179*, 292–307. [CrossRef.].
95. Mohajerani, A., Vajna, J., Cheung, T. H. H., Kurmus, H., Arulrajah, A., & Horpibulsuk, S., (2017). Practical recycling applications of crushed waste glass in construction materials: A review. *Constr. Build. Mater., 156*, 443–467. [CrossRef.].
96. Morgan, P. W., (1979). *Aromatic Polyamides, 1*, 1–29.
97. Murday, J. S., Dominguez, D. D., Moran, Jr. J. A., Lee, W. D., & Eaton, R., (1984). An assessment of graphitized carbon fiber use for electrical power transmission. *Synthetic Metals, 9*(3), 397–424.
98. Myadaraboina, H., Law, D., & Patnaikuni, I., (2014). Durability of basalt fibers in concrete medium. In: *Australasia and Southeast Asia Conference in Structural Engineering and Construction (ASEA-SEC-2)* (Vol. 2, pp. 1–22). ISEC Press: Fargo, ND, USA.
99. Namboonruang, W., & Yongam-Nuai, P., (2016). Thermal, strength and leachability characteristics of cellulose fiber-reinforced lime-soil brick. *Key Eng. Mater., 703*, 386–395. [CrossRef.].

100. Northolt, M. G., (1981). Compressive strength and glass transition temperature. *Journal of Materials Science, 16*(7), 2025–2028.

101. Panda, B., Paul, S. C., & Tan, M. J., (2017). Anisotropic mechanical performance of 3D printed fiber-reinforced sustainable construction material. *Mater. Lett., 209*, 146–149. [CrossRef.].

102. Parkyn, B., (1970). *Glass Reinforced Plastics (Papers on Glass Fiber Reinforced Plastics Covering End Products and design Criteria), 1*, 1–20.

103. Pelisser, F., Montedo, O. R. K., Gleize, P. J. P., & Roman, H. R., (2012). Mechanical properties of recycled PET fibers in concrete. *Mater. Res., 15*, 679–686. [CrossRef.].

104. Perret, R., & Ruland, W., (1970). The microstructure of PAN-base carbon fibers. *Journal of Applied Crystallography, 3*(6), 525–532.

105. Petková, M., Hýlová, M., & Ujhelyiová, A., (2017). Physical modification of polypropylene fibers used in construction. (2017). *J. Text. Inst., 108*, 196–202.

106. Petrovic, J. J., Milewski, J. V., Rohr, D. L., & Gac, F. D., (1985). Tensile mechanical properties of SiC whiskers. *Journal of Materials Science, 20*(4), 1167–1177.

107. Pollack, J. T. A., (1972). Growth of void-free sapphire filaments. *J. Mater. Sci, 7*, 787–792.

108. Putman, B. J., & Amirkhanian, S. N., (2004). Utilization of waste fibers in stone matrix asphalt mixtures. *Resour. Conserv. Recycl., 42*, 265–274. [CrossRef.].

109. Raut, A. N., & Gomez, C. P., (2017). Development of thermally efficient fiber-based eco-friendly brick reusing locally available waste materials. *Constr. Build. Mater., 133*, 275–284. [CrossRef.].

110. Riggs, J. P., (1985). Casein. *Encyclopedia of Polymer Science and Engineering, 2*, 685–691.

111. Roesler, J. R., Altoubat, S. A., Lange, D. A., Rieder, K. A., & Ulreich, G. R., (2006). Effect of synthetic fibers on structural behavior of concrete slabs-on-ground. *Mater. J., 103*, 3–10.

112. Saad, S., Mirzababaei, M., Mohamed, M., & Miraftab, M., (2012). Uniformity of density of compared fiber reinforced clay soil samples prepared by static compaction. In: *Proceedings of the 5th European Geosynthetics Congress* (Vol. 5, pp. 1–33). Valencia, Spain.

113. Saha, P., Chowdhury, S., Roy, D., Adhikari, B., Kim, J. K., & Thomas, S., (2016). A brief review on the chemical modifications of lignocellulosic fibers for durable engineering composites. *Polym. Bull., 73*, 587–620. [CrossRef.].
114. Sakka, S., (1985). Sol-gel synthesis of glasses: Present and future. *American Ceramic Society Bulletin, 64*(11), 1463–1466.
115. Serin, S., Morova, N., Saltan, M., & Terzi, S., (2012). Investigation of usability of steel fibers in asphalt concrete mixtures. *Constr. Build. Mater., 36*, 238–244. [CrossRef.].
116. Shindo, A., (1961). Studies on graphite fiber. *Report of Osaka Industrial Technology Laboratory*, (317), 1–50.
117. Siddika, A., Al Mamun, M. A., Alyousef, R., & Amran, Y. M., (2019). Strengthening of reinforced concrete beams by using fiber-reinforced polymer composites: A review. *J. Build. Engineering, 25*, 100798. [CrossRef.].
118. Simon, G., & Bunsell, A. R., (1984). Mechanical and structural characterization of the Nicalon silicon carbide fiber. *Journal of Materials Science, 19*(11), 3649–3657.
119. Singer, L., (1979). High modulus carbon fibers from. *Ultra-High Modulus Polymers,* 1, 240–251.
120. Smith, P., Lemstra, P. J., Kalb, B., & Pennings, A. J., (1979). Ultrahigh-strength polyethylene filaments by solution spinning and hot drawing. *Polymer Bulletin, 1*(11), 733–736.
121. Soares, B., Preto, R., Sousa, L., & Reis, L., (2016). Mechanical behavior of basalt fibers in a basalt-UP composite. *Procedia Struct. Integr., 1*, 82–89. [CrossRef.].
122. Soltani, A., Deng, A., Taheri, A., & Mirzababaei, M., (2017). A sulfonated oil for stabilization of expansive soils. *Int. J. Pavement Eng.,* 1–14. [CrossRef.].
123. Suresh, S., & Kumar, V. S., (2014). Experimental determination of the mechanical behavior of glass fiber reinforced polypropylene composites. *Procedia Eng., 97*, 632–641. [CrossRef.].
124. Svedberg, R., Vitale, M., Laine, R. M., Chew, K. W., & Weinmann, M., (1998). *Low Cost, YAG and Mullite Fibers by Continuous Extrusion a Pyrolysis of Metal Carboxylate Precursors* (Vol. 1, pp. 1–22). Advanced Modular Power Systems Inc Ann Arbor Mi.

125. Wang, M. C., Zhang, Z. G., Li, Y. B., Li, M., & Sun, Z. H., (2008). Chemical durability and mechanical properties of alkali-proof basalt fiber and its reinforced epoxy composites. *J. Reinf. Plast. Compos., 27*, 393–407. [CrossRef.].

126. Xue, Q., Feng, X. T., Liu, L., Chen, Y. J., & Liu, X. L., (2013). Evaluation of pavement straw composite fiber on SMA pavement performances. *Constr. Build. Mater., 41*, 834–843. [CrossRef.].

127. Yang, H. H., (1993). In: Seymour, R. B., & Porter, R. S., (eds.), *Manmade Fibers: Their Origin and Development* (Vol. 1, pp. 1–29).

128. Yin, S., Tuladhar, R., Shi, F., Combe, M., Collister, T., & Sivakugan, N., (2015). Use of macro plastic fibers in concrete: A review. *Constr. Build. Mater., 93*, 180–188. [CrossRef.].

Chapter

2

Fundamentals of Polymers

CONTENTS

2.1. Introduction 48

2.2. The History of the Concept of the Macromolecule 49

2.3. Classification of Polymers 51

2.4. Structure and Properties of Polymers 53

2.5. Thermoplastic Polymers 53

2.6. Thermosetting Polymers 61

2.7. Naturally Occurring Polymers 68

References 73

2.1. INTRODUCTION

A polymer is a large molecule made up of several smaller molecules. These larger molecules may be highly interconnected, slightly brancher, or linear. In the former case, the structure forms into a large three-dimensional network (Morgan and Gilman, 2013; Dhote et al., 2019). Monomers are the small molecules that are used as the basic building blocks of these large molecules. For instance, the commercially significant material poly(vinyl chloride) is formed from the monomer vinyl chloride. The replication unit in the polymer typically resembles the monomer from which the polymer was prepared. There are exemptions to this, however. Poly(vinyl alcohol) is properly taken into account to be made up of vinyl alcohol (CH_2CHOH) repeat units, but there is, in detail, no such monomer as vinyl alcohol. In the alternative tautomeric form ethanal CH_3CHO, the suitable molecular unit occurs. It is essential first to make poly(vinyl ethanoate), to make this polymer, from the monomer vinyl ethanoate, and then to hydrolyze the product to produce the polymeric alcohol (Bloembergen, 1996; Roth and Baglay, 2016).

The number of repeat units and mass of polymer molecules determines the size of the polymer molecules in the molecule. The former indicator of the size is known as the *degree of polymerization,* DP. Thus, the product of the relative molar mass of the DP and the repeat unit is known as the relative molar mass of the polymer. There is no definite boundary between the rest of chemistry and polymer chemistry. As a very vague estimation, molecules of a DP of at least 100 or a relative molar mass of at least 1000 are taken to fall into the field of polymer chemistry (Helfand, 1975; Chandrasekhar, 1999).

Several polymers in commercial usage are organic, as they are based on the covalent compounds of carbon. This is also valid for the silicones, which also usually comprise substantial proportions of a hydrocarbon group, though based on silicon-oxygen backbones. The other elements included in polymer chemistry most frequently are chlorine, phosphorus, oxygen, sulfur, and fluorine. Those elements which are capable of forming covalent bonds, albeit of the same polarity, with carbon (Figure 2.1) (Migliaresi and Pegoretti, 2002; Prakash et al., 2015).

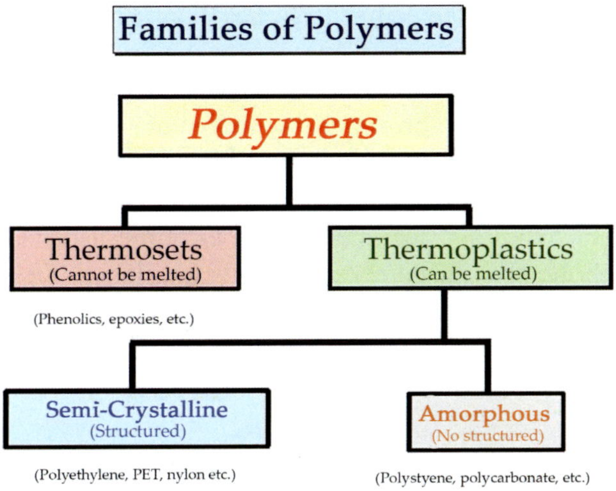

Figure 2.1. Classification of polymers.

Source: https://www.azom.com/article.aspx?ArticleID=17477.

2.2. THE HISTORY OF THE CONCEPT OF THE MACROMOLECULE

The word polymer is derivative of the Greek words 'poly' meaning many and 'meros' meaning part as explained in the modern books about polymer chemistry. They frequently then conclude that it follows that this term relates to massive molecules built up of large numbers of interrelated monomer units (Arturson, 1979; Minale et al., 1997). This is deceptive since factually the word polymer was devised for other reasons. The idea of polymerism was initially applied to the condition in which molecules had very different physical and chemical properties but tical empirical formulae. For instance, benzene (C_6H_6; empirical formula CH) was thought to be a polymer of acetylene (C_2H_2; empirical formula also CH). Thus, the word 'polymer' is to be present in textbooks of organic chemistry available up to about 1920 but not with its contemporary meaning (Tobolsky and Eyring, 1943; Covas et al., 2013).

However, in the case of some chemicals, the situation is mixed up. For instance, styrene was identified from the mid-19th century as a clear organic liquid of distinctive pungent smell. It was also recognized to change itself under definite conditions into a clear resinous solid that was almost odor-free, this resin then being named metastyrene (Wohrle and Pomogailo, 2003). The polymerization was termed as a process of the formation of metastyrene from styrene and metastyrene was detained to be a polymer of styrene. However, despite the very profound alteration of physical and chemical properties, these expressions applied only in the sense that there was no alteration in the empirical formula. There was no mention of the reason for this change, and definitely, the chemists of the time had no clue of what had occurred to the styrene that was distantly similar to the contemporary view of polymerization (Figure 2.2) (Muñoz-Bonilla and Fernández-García, 2012).

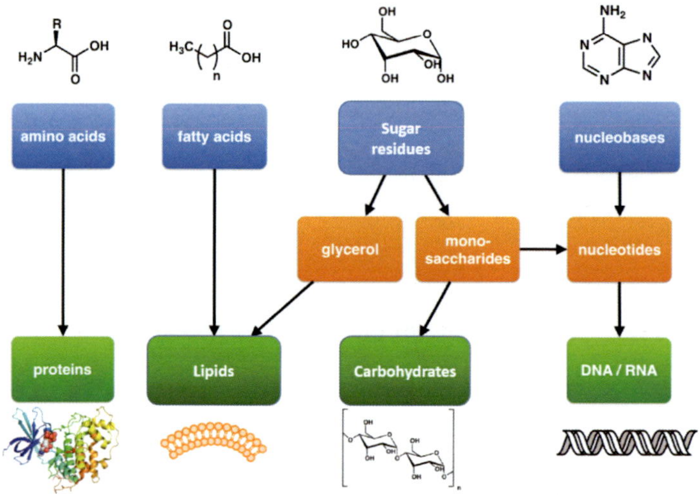

Figure 2.2. Different types of macromolecules.

Source: https://wou.edu/chemistry/chapter-11-introduction-major-macromolecules/.

Understanding of the essential nature of those materials now named polymers had to pause until the 1920s when Herman Staudinger clarified thinking after devising the word 'macromolecule.' There was no uncertainty about this new word, which these days is used almost interchangeably with the word polymer, -it meant 'large molecule,' again from the Greek. Strictly speaking, however, the words are not synonymous (Abir-Am, 1992; Karplus, 2003). There is no intention in principle for a macromolecule to be made up of *repeating* structural units; in usage, however, they typically are. Staudinger's idea of macromolecules was not at all well established at first. His wife once recollected that he had 'met disagreement in all his lectures.' The characteristic of this disagreement was that of one famous organic chemist who acknowledged that it was as if zoologists 'were told that somewhere in Africa, an elephant was found who was 1500 feet long and 300 feet high (Williams, 1984; Wu et al., 2008).

2.3. CLASSIFICATION OF POLYMERS

There are various methods of classification of polymers. The study of their response to the thermal treatment is one of the approaches followed by their division into thermosets and thermoplastics. Thermosets are polymers that do not melt when heated but decompose irreversibly at sufficiently high temperatures, while thermoplastics are polymers that melt when heated and resolidify when cooled (Doi and Edwards, 1978; Seymour, 1992). This system has the advantage that between the two groups, there is a valuable chemical difference. Thermosets are considerably crosslinked materials, containing a widespread three-dimensional network of covalent chemical bonding, while thermoplastics contain fundamentally lightly branched or linear polymer molecules (Figure 2.3) (Leitner et al., 2003; Anzano et al., 2008).

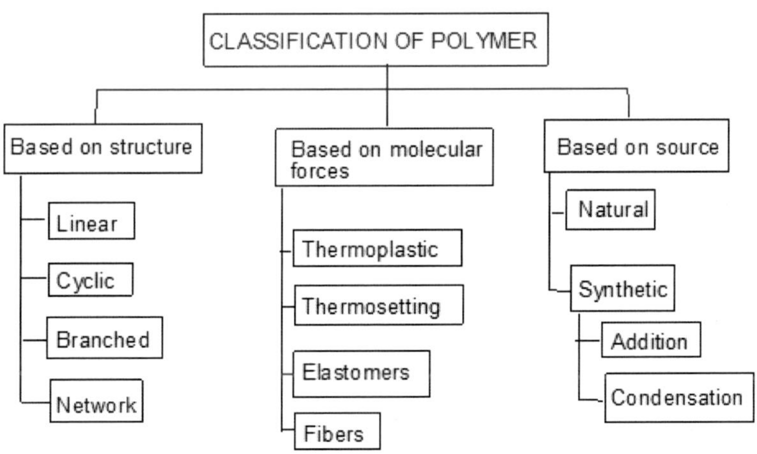

Figure 2.3. Detailed classification of polymers.

Source: https://www.toppr.com/ask/content/concept/advanced-knowledge-of-polymerization-203615/.

Another system of classification is based on the nature of the chemical reactions engaged in polymerization, first recommended by Carothers in

1929. Here the two main groups are the *addition* and the *condensation* polymers. Condensation polymers are those made from monomers where the reaction is supplemented by the loss of a small molecule, type of water, for instance, polyesters which are made by the condensation displayed in Reaction 1.1 (Papkov, 1982; Kulcke et al., 2003).

$$n\,HO{-}R{-}OH \;+\; n\,HOOC{-}R^1{-}COOH \longrightarrow HO[{-}R{-}COO{-}R^1{-}COO{-}]_nH \;+\; (n-1)H_2O \qquad (1.1)$$

Addition polymers, on the other hand, are those made by the addition reaction of an unsaturated monomer, such as occurs in the polymerization of vinyl chloride (Reaction 1.2).

$$n\,CH_2{=}CHCl \longrightarrow [{-}CH_2{-}CHCl{-}]_n \qquad (1.2)$$

P. J. Flory, to some extent modified this system, the importance was placed on the mechanisms of the polymerization reactions. He regrouped the polymerizations as *chain* reactions or *step* reactions conforming nearly to addition or condensation in Carothers scheme, but not entirely. A distinguished exemption takes place with the formation of polyurethanes (PURs), which are made by reaction of isocyanates with hydroxy compounds and follow 'step' kinetics, but without the removal of a small molecule from the particular units (Reaction 1.3).

$$n\,OCN{-}R{-}NCO \;+\; n\,HO{-}R^1{-}OH \longrightarrow OCN{-}[R{-}NHCO_2{-}R^1{-}]_n\,OH \qquad (1.3)$$

In the leading of these, the kinetics are such that there is a steady build-up of high relative molar mass material as the reaction moves, with the highest molar mass molecules not showing until the very end of the reaction. Contrary to this, chain reactions, which happen only at a comparatively few activated sites within the reacting medium, happen with the fast build-up of insufficient high relative molar mass molecules while the rest of the monomer keep unreacted. When made, such macromolecules stay fundamentally unaffected while the rest of the monomer experiences alteration. This means that large molecules show very early in the polymerization reaction, which is characterized by having both high relative molar mass and monomer molecules existing for most of the period of the reaction (Boeykens and Vázquez, 2004).

2.4. STRUCTURE AND PROPERTIES OF POLYMERS

After establishing the basic rules of classification in polymer chemistry, now attention will be given to individual polymers, particularly on their chemical and physical properties. Several examples which follow are of commercial significance, though it is their properties that are given importance rather than the whole chemistry of their industrial production. There are two main classifications of polymers, namely thermoplastics and thermosets. In terms of the classification just mentioned, it is also found that the polymers of main commercial significance are thermoplastics or addition polymers rather than thermosets and these will be discussed first (Ringsdorf, 1975; Garbarczyk and Paukszta, 1985).

2.5. THERMOPLASTIC POLYMERS

In this section, various thermoplastic polymers are discussed in detail (Novikov and Kozlov, 2000).

2.5.1. Poly(Ethylene)

One of the simplest molecular structures is present in these polymers ($[CH_2CH_2-]_n$) and is currently the largest tonnage plastic material, having first been made commercially in 1939 for its use in electrical insulation. The nomenclature of this polymer encounters difficulty. Ethane is the IUPAC recommended name for the monomer, rather than the previous ethylene. Hence poly(ethene) is the IUPAC name for the polymer (Aranda and Ruiz-Hitzky, 1992; Israelachvili, 1997).

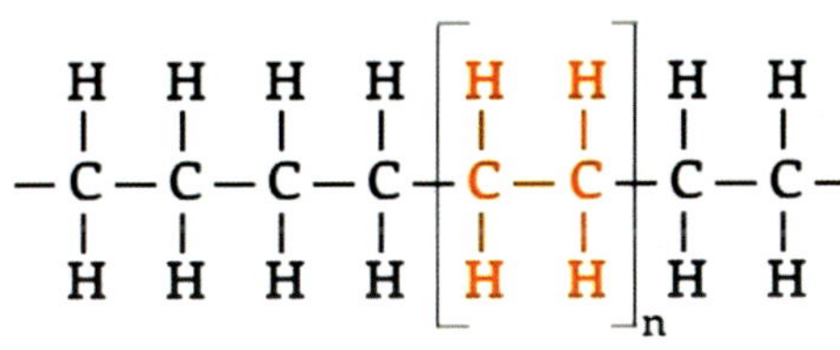

Molecular Structure of Polyethylene

Figure 2.4. The structural formula of polythene.

Source: https://omnexus.specialchem.com/selection-guide/polyethylene-plastic.

However, chemists have never used this name when working with the material; therefore; throughout this book, this polymer will be denoted by its more extensive name, poly(ethylene) (Figure 2.4).

For the preparation of poly(ethylene), there are four different industrial methods, and they produce products that have somewhat different properties. The four methods are (Nucci et al., 1991; Busico and Cipullo, 2001):

- Ziegler processes;
- High-pressure processes;
- The standard oil (Indiana) process; and
- The Phillips processes.

The basis of the Ziegler processes is on coordination reactions catalyzed by metal alkyl systems. Karl Ziegler in Germany discovered such reactions and was developed in the early 1950s by G. Natta at Milan.

A complex prepared from the triethylaluminum and titanium tetrachloride is a typical Ziegler-Natta catalyst. First, it is fed into the reaction vessel, ethylene is added afterward. Low temperatures, typically no more than 70°C and low pressures are usually required for this type of reaction with the rigorous exclusion of moisture and air, which would damage the catalyst. Such processes produce poly(ethylene) which are of intermediate density, having a value of about 0.945 g cm^{-3}. A relative molar mass may be attained for such polymers by changing the ratio of the catalyst components or by presenting a small quantity of hydrogen into the reaction vessel (Padden and Keith, 1959).

High-pressure processes use a temperature of 80–300°C and pressure of 1000–3000 atm. Free-radical initiators, such as oxygen or benzoyl peroxide are usually used, and to prevent a runaway reaction conditions need to be carefully controlled, which would produce methane, graphite, and hydrogen rather than polymer. Overall, high-pressure processes incline to produce lower density poly(ethylenes), naturally in the range 0.915–0.945 g cm^{-3}, which also have comparatively low molar masses.

Lastly, the Standard Oil (Indiana) and the Phillips Processes both produce high yield high-density poly(ethylenes), using comparatively low temperatures and pressures (Jones et al., 1964). In the modern world, the polymer is very familiar material. It is a wax-type solid which is easily processed, relatively low in cost, and shows good chemical resistance. Environmental stress cracking may occur in low relative molar mass grades, that is, they abruptly fail catastrophically for no obvious reason after exposure to moisture and sunlight. Despite this problem, the numerous grades of poly(ethylene) have an extensive range of applications. These include components for chemical plants, s for electrical insulation, packaging, crates, and pipes.

2.5.2. Poly(Propylene)

This polymer having the structure [—$CH_2CH(CH_3)$—]$_n$ come up as a marketable material succeeding the work of Natta on catalysts for the formation of high relative molar mass polymers from alkenes. Succeeding his work on the polymerization of ethylene, Natta displayed in 1954 that it was likely to make similar polymers of propylene. Commercial manipulation followed quickly, and poly(propylene) was first commercialized in 1957 (Figure 2.5) (Jones et al., 1964).

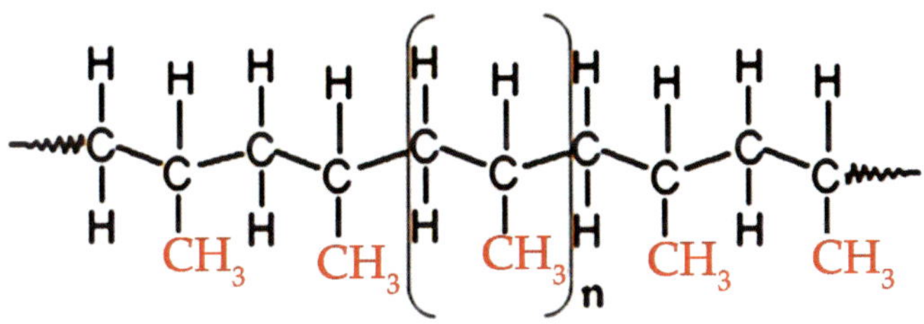

Figure 2.5. The structural formula of polypropylene.

Source: https://omnexus.specialchem.com/selection-guide/polypropylene-pp-plastic.

There are recognized issues with the nomenclature of this polymer like poly(ethylene), since its IUPAC name, poly(propene), is also seldom if ever used by polymer chemists. Since, in actual practice, no uncertainty is linked with the non-systematic name, this is the one that is normally used, as it will be all over this book (Padden and Keith, 1959).

Poly(propylene) was found to exist in two possible forms when it was first made. One was analogs to poly(ethylene), but had greater hardness and rigidity; the other was found to be of little strength and amorphous. The first of these is now identified to be isotactic, that is at each alternating carbon atom with a regular stereochemistry. The other is now recognized to be atactic, that is at each methyl-bearing carbon atom with a random distribution of different stereochemical arrangements (Table 2.1) (Busico and Cipullo, 2001).

Table 2.1. Details of Processes for Preparing Poly(Ethylene)

Process	Catalyst	Pressure (atm)	Temperature (°C)	The Density of Product (g cm^{-3})
Standard Oil (Indiana)	Supported MoO_3 with Na, Ca metal, or hydride promoters	40–80	230–270	9.6
Phillips	5% CrO_3 in finely divided silica/ alumina	15–35	130–160	9.6

Commercially poly(propylene) is typically about 90–95% isotactic and is very analogous to poly(ethylene) with the following exclusions:

- It can be used at high temperatures because of its higher softening point. For instance, it is used to make kettles, usually of the jug-type of design, and it is observed to be well able to endure the impacts of exposure to boiling water.
- It has a lower density (0.90 g cm^{-3}).
- It is more quickly oxidized. This is a result of the comparatively readily fragmented tertiary C—H bond within the molecule.
- It is not vulnerable to environmental stress cracking. Another exciting property of isotactic poly(propylene) is that it can be recurrently flexed without causing weakness. Hence it is used for one-piece moldings, such as boxes for card indexes, which can be used over many years without harm.

2.5.3. Poly(Methyl Methacrylate)

The most vital of the commercial *acrylic* polymers is the polymer whose name is frequently abbreviated to PMMA. These polymers are properly derived from poly(acrylic acid), [—$CH_2CH(CO_2H)$—]$_n$. In the situation of PMMA, this derivation can be assumed as coming about by substitute of the tertiary hydrogen atom by a methyl group, CH_3, and by esterification of the carboxylic acid group with methanol, CH_3OH. This gives the succeeding structure (Figure 2.6) (Wu, 1970; Jia et al., 1999).

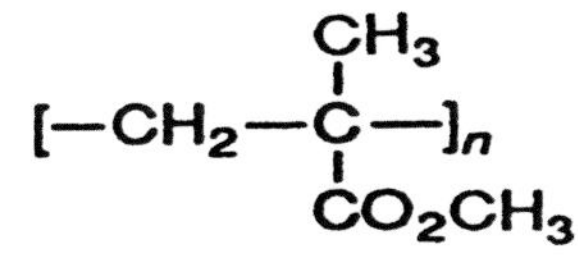

(1.1)

Figure 2.6. The structural formula of PMMA polymer.

Source: https://www.degruyter.com/document/doi/10.1515/pac-2016-0502/html.

The IUPAC-recommended name for PMMA is seldom used, as with other main tonnage polymers, partially no doubt because it is tremendously cumbersome. It is poly[l-(methoxycarbonyl)-l-methylethene].

As time passes, PMMA is becoming comparatively less important commercially. For instance, by 1980, the sales of the PMMA had fallen to 12%, which ran at 40% of the total level of sales of poly(styrene) in the mid-1960s. PMMA does interestingly as the material from which plastic dentures are produced, though it retains certain uses (Russell et al., 1990).

PMMA is a glassy, transparent material that is identified as amorphous at the molecular level from X-ray studies. It is polar, as is revealed by its solubility in solvents such as trichloroethane and propanoate and its or shielding properties comparative to poly(ethylene).

The major applications for PMMA come from the coupling of its good outdoor weathering properties and transparency. These are combined with appropriate rigidity and toughness so that PMMA is a suitable material in a range of glazing applications. For instance, it is used for street lamp fittings, display signs, and ceiling lights in offices and factories. It is also standard material for automobile rear lamp housings (Ali et al., 2015).

2.5.4. Poly(Styrene)

In the developed world, a polymer named poly(styrene) finds widespread based on its required properties, coupled with its relative cheapness. Rigidity, excellent color range, low water absorption, and transparency are among their distinctive characteristics. The styrene monomer is derived from vinyl benzene. It is a mobile liquid having colorless features that polymerize readily. The first shot of the polymerization reaction came in 1839 when E. Simon explained the transformation of what was then named 'styrol.' He assumed he had oxidized the material and named the product styrol oxide. Afterward, when it was recognized that it had no oxygen, the product became called metastyrene (Figure 2.7) (Cornelisse et al., 1998; Prucker and Rühe, 1998).

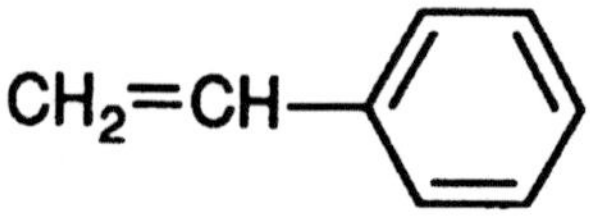

(1.2)

Figure 2.7. Structure of vinyl benzene (styrene).

Source: https://en.wikipedia.org/wiki/Styrene.

The basis of the polymer is the simple head-to-tail structure of monomer units and is amorphous since the precise location of the benzene ring is slightly flexible and therefore hinders crystallization. It is considered too brittle for many applications despite its largely desirable properties. Various approaches have been made to modify the mechanical properties of poly(styrene) due to the above-mentioned reason. The most acceptable of these have been: (i) the addition of rubbery fillers; and (ii) copolymerization.

Numerous copolymers of styrene have been developed and sold commercially. Polymers containing butadiene are among the most successful of those polymers. These copolymers act as synthetic elastomers at the suitable ratio of monomers, called styrene-butadiene rubber, SBR. Terpolymers of butadiene, acrylonitrile, and styrene, ABS are another significant group of materials. A large number of ABS materials can be formed by changing not only the ratios of the three monomers but also the exact circumstances of manufacture. ABS materials usually have high impact strength and can be readily molded to give objects of good surface finish (Wilhelm et al., 1991).

The use of rubbery fillers is the alternative method to produce strengthened styrene polymers. Such type of materials is called high-impact poly(styrenes), a name which shows their enhanced mechanical properties. They are naturally made by softening the rubber in the styrene monomer, followed by polymerization of the styrene in the normal mode. The method gives a blend comprising not only rubber and poly(styrene) but also small quantities of the graft copolymer, containing short poly(styrene) side chains linked to the molecules of the rubber. The general result of using this method is to give materials with much better impact strengths than pure poly(styrene).

Lastly, for crosslinking unsaturated polyesters, the polymerization of styrene is used. Normally, a polyester is formed which holds double bonds. Such a resin, called a pre-polymer, is then mixed with styrene to give a gummy solution, and this solution is then used into a glass fiber mat. The resultant composite material establishes its ultimate properties by polymerization of the styrene, which fundamentally experiences homopolymerization, but also includes the random double bond from the pre-polymer molecule. This gives a large, three-dimensional linkage, intertwined with the glass strengthening. The composite material has outstanding properties and is extensively used for boat hulls, baths, and shower units.

2.5.5. Poly(Vinyl Chloride) (PVC)

For poly(chloroethene), the widely accepted trivial name is poly(vinyl chloride) and is one of the three most significant polymers in present use, in terms of worldwide production, the other two being poly(styrene) and poly(ethylene). PVC has several uses, counting packaging, cable insulation, and toys (Figure 2.8) (Xu and Xu, 2002).

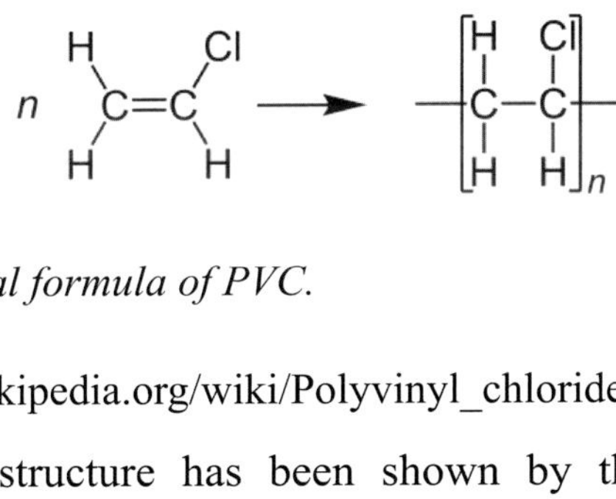

Figure 2.8. *Structural formula of PVC.*

Source: https://en.wikipedia.org/wiki/Polyvinyl_chloride.

A head-to-tail structure has been shown by the PVC. Distinctive experimental indication for this is that when treated with zinc dust and dissolved in dioxin. Distinctive experimental indication for this is that when dissolved in dioxane and treated with zinc dust, it experiences a Wurtz-type reaction to produce a product having a small quantity of chlorine and no measurable unsaturation (Schurer et al., 1975; Braun, 2004). The substitute-likely structure, the head-to-head structure, would produce unsaturated sites where nearby chlorine atoms had been detached (Reaction 1.4).

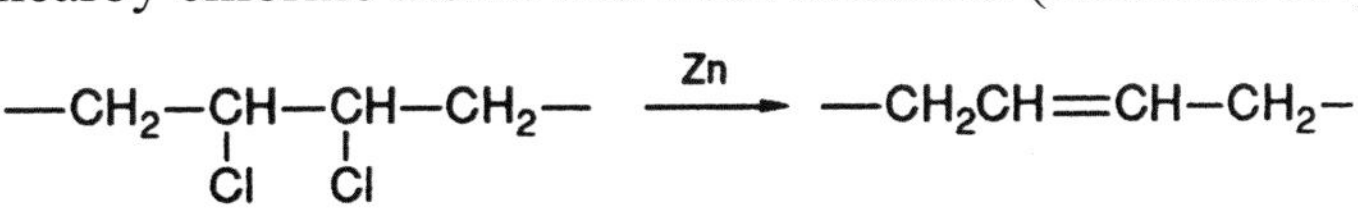

(1.4)

2.5.6. The Nylons

Du Pont Company of America gave the name nylon to their first synthetic condensation polymer formed by the reaction of difunctional amines with difunctional acids. To examine the whole topic of polymerization, it had been made as part of the fundamental program of W. H. Carothers. The word has progressively been protracted to other associated polymers. These materials are firmly polyamides (PAs), but this term contains that otherwise separate category of natural macromolecules, the proteins. The term nylon is reserved for its practicality in unique synthetic PAs from the wider category of such polymers (Kang and Chung, 2000).

As mentioned above, nylons are step or condensation polymers, and because of this, they are dissimilar from all the other commercially vital polymers termed so far. The nylons are separated from each other by a numbering system based on the number of carbon atoms in the starting materials. Thus, nylon 6,6, which was first formed in 1935 and is still the chief marketable nylon, is formed by the reaction of adipic acid (six carbon atoms) with hexamethylenediamine (six carbon atoms) (Reaction 1.5).

$$n\,H_2N(CH_2)_6NH_2 \;+\; n\,HOOC(CH_2)_4COOH \longrightarrow$$

$$HN_2[(CH_2)_6NHCO(CH_2)_4CO]_n \;+\; (n-1)H_2O \qquad (1.5)$$

In the same way, nylon 6,10 is formed from sebacic acid (ten carbon atoms) and hexamethylenediamine. Some nylons are selected not by two numbers, but by a single number; these are the nylons formed from a distinct starting material. Such a starting material can be either a closed ring amide-type compound, called a lactam, which will experience a ring-opening reaction to produce a polymer or a long-chain amino acid, such as ω-aminodecanoic acid (11 carbon atoms) which experiences self-condensation to produce nylon 11 (Zachariades and Porter, 1979). This is the process of making nylon 6 from caprolactam (six carbon atoms) and nylon 12 is formed from dodecyl-lactam (12 carbon atoms). The second reaction may be signified by the equation displayed in Reaction 1.6.

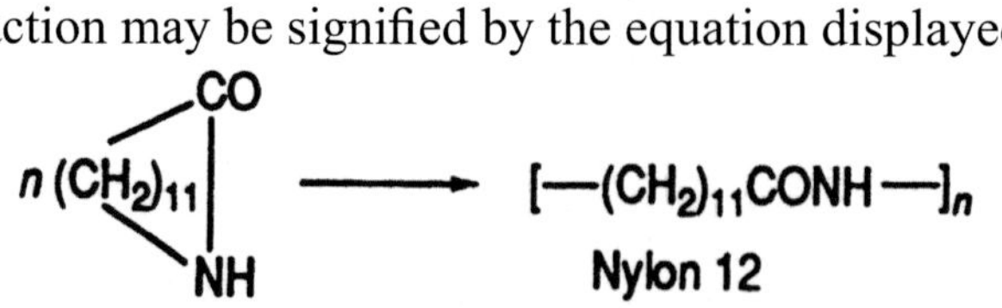

(1.6)

Several nylons have similar physical properties. These include flexibility, high impact strength, abrasion resistance, and toughness. Also, they form outstanding fibers because of the linear structure. The early commercial success of the nylon 6,6 was based on these above factors. In

1938, commercial production of fibrous nylon 6,6 began and 2 years later, nylon stockings became normally accessible in the USA. They were a huge success as is well known and became extremely fashionable during the years of World War II. Organic solvents have very little effect on nylon and oils and fuels have little or no impact on it. However, they are readily attacked by alkalis at elevated temperatures and concentrated mineral acids at room temperature (Jin et al., 2013).

Usually, Nylon 6,6, and 6 are employed as textile fibers. Where separate monofilaments are used, such as in sports equipment, brushes, or surgical sutures, nylon 6, 10, and 11 are usually used, because of their greater water resistance and flexibility. All nylons can be injection molded, and the resultant objects have found extensive use in engineering applications, such as gears and bearings.

2.6. THERMOSETTING POLYMERS

In this section, various thermosetting polymers are discussed in detail (Yamini and Young, 1980).

2.6.1. Epoxy Resins

The three-membered heterocyclic system is the basis of these polymers either as oxirane or epoxy ring (Figure 2.9).

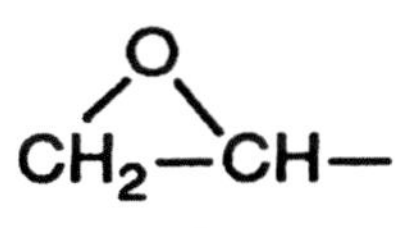

(1.3)

Figure 2.9. The structural formula of epoxy.

Source: https://data.epo.org/publication-server/document?iDocId=363171&iFormat=2.

Diepoxides such as that derivative of bisphenol A are engaged, for commercial application, and their curing takes place via ring-opening crosslinking reactions, into which the epoxy groups enter readily. It is called Bisphenol A because it is prepared to form two moles of acetone and phenol (Reaction 1.7).

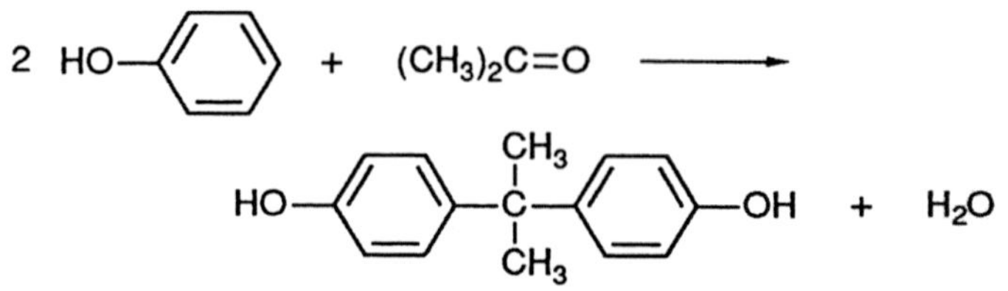

(1.7)

These ring openings can be brought about by a variety of reagents, including anhydrides and amines used in stoichiometric amounts, or Lewis bases such as tertiary amines and Lewis acids such as $SnCl_{4,}$ used in catalytic amounts.

Reaction 1.8 shows the crosslinking reactions, and they validate that, only a small amount of curing agent is required to bring about the cure of epoxy resins. Several factors are taken into account in the selection of the curing agent, such as ease of handling, cure rates, cost, pot life, and the electrical, mechanical, and thermal properties essential in the final resin.

Comparatively, epoxy resins are expensive, but despite this, they are inflexibly developed in several significant applications. These comprise protective coatings, adhesives, laminates, and a variety of uses in construction and building (Bucknall and Gilbert, 1989).

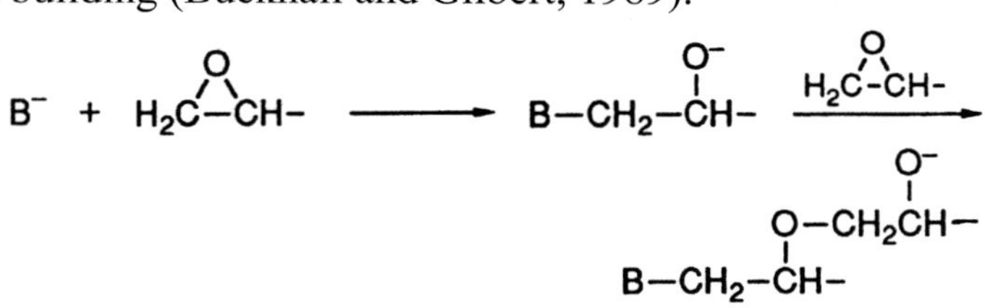

(1.8)

2.6.2. Phenol-Formaldehyde Polymers

These polymers are thermoset polymers prepared from phenol or, less frequently phenolic (PF) type compounds such as resorcinol, cresols, and xylenols, coupled with formaldehyde. G.T. (later Sir Gilbert) Morgan discovered them in the early 1890s when trying (unsuccessfully) to make synthetic dyestuffs by reaction of formaldehyde with phenol and they had been known for some time. But before 1907, the knowledge had not been exploited the year in which Leo Baekeland obtained his first patent in America for materials made from these two compounds. The General Bakelite Company was founded in 1910 to exploit this development, in the process of making phenol-formaldehyde, the first artificial polymers to attain commercial significance (Figure 2.10) (Costa et al., 1997).

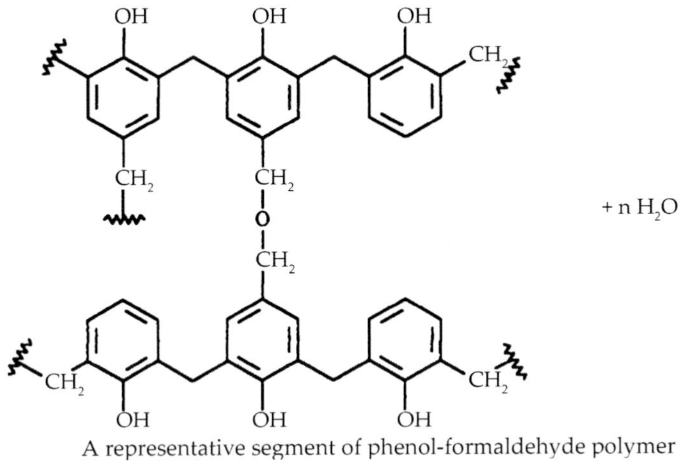

Figure 2.10. The segment of the phenol-formaldehyde polymer.

Source: https://www.sciencedirect.com/topics/earth-and-planetary-sciences/phenol-formaldehyde.

Before he could bridge the gap between the initial concept and final products, Baekeland had to make significant discoveries. Specifically, he observed that the base-catalyzed condensation of formaldehyde and phenol can take place in two parts. An intermediate product can be isolated if the process is carefully controlled, either as a solid or a liquid, depending on the degree of the reaction. At this step, the material contains fundamentally linear molecules and is both soluble and fusible in suitable solvents. This intermediate is converted to the infusible, hard solid known as 'Bakelite' when heated under pressure to 150°C. The three-dimensional cross-linked network develops at the second step (Poljansek and Krajnc, 2005).

Where synthetic polymers are concerned, phenol-formaldehyde may no longer hold the center stage, but they are still of some commercial significance. They are formed for electrical moldings, household fittings, appliance handles, and as specialized surface coatings and adhesives.

Cured phenol-formaldehyde is resilient to the outbreak by most chemicals. Water and organic solvents have no impact on them, though swelling is observed in boiling phenols. Sodium hydroxide solutions can attack simple resins readily, but these solutions have the least effect on the resins based on phenol derivatives, such as cresol. Most acids have no effect on simple phenol-formaldehyde polymers, though nitric and formic acids will tend to attack them. Again, such attacks face resistance to the cresol-based polymers (Moubarik et al., 2009; El Mansouri et al., 2011).

2.6.3. Amino Resins

Polymers that are prepared by the reaction of either melamine or urea with formaldehyde are termed amino resins. In both these reactions, the product formed has a well-crosslinked network structure and therefore is a thermoset polymer. Figure 2.11 shows the structure of the two-parent amino compounds (Nguon et al., 2018).

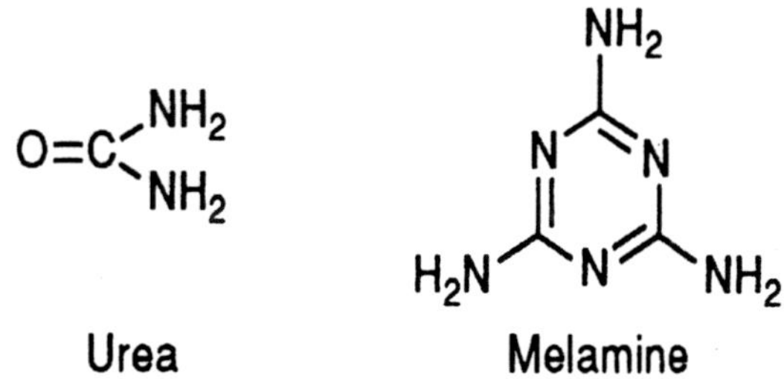

Figure 2.11. Structures of urea and melamine.

Source: https://www.sciencedirect.com/topics/engineering/melamine-urea-formaldehyde-resin.

The reaction of the amine groups with formaldehyde is the initial step of the polymerization process formaldehyde to generate methylol units, as shown in Reaction 1.9. A variety of reactions result from further heating of the polymer. For instance, self-condensation of the methylol groups can take place (Reaction 1.10).

$$-NH_2 + HCHO \longrightarrow -NHCH_2OH \quad (1.9)$$

$$2\ -NHCH_2OH \longrightarrow -NHCH_2OCH_2NH- + H_2O \quad (1.10)$$

On the other hand, the methylol groups can react with more amino groups, also developing a molecule of water in what is alternative condensation reaction (Reaction 1.11).

$$-NHCH_2OH + H_2N- \longrightarrow -NHCH_2NH- + H_2O \quad (1.11)$$

The amino resins are not themselves intensely colored, unlike phenol-formaldehyde polymers, but are of a natural light appearance. They can be simply pigmented to produce a diversity of shades, which takes it to application in uses where decent appearance is extremely appreciated, for instance, in laminated resins for furniture, decorative tableware, and modern white electrical sockets and plugs (Gao et al., 2004; Torlakoğlu and Güçlü, 2009).

Amino resins are very resilient to most organic solvents when crosslinked, though they tend to be attacked by both alkalis and acids. Urea-formaldehyde polymers are more vulnerable to attack than those prepared from formaldehyde and melamine (Abd El-Magied et al., 2016).

2.6.4. Poly(Tetrafluoroethylene) (PTFE)

This polymer is the totally fluorine-substituted equivalent of poly(ethylene), i.e., $[CF_2CF_2–]_n$. The quantity of this polymer formed commercially is very minor equated with the yield of numerous other synthetic polymers, but it has several significant specialized applications and so is worth seeing briefly (Figure 2.12) (Rae and Dattelbaum, 2004).

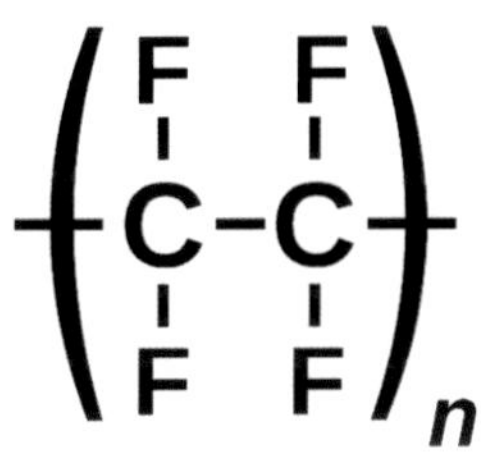

Figure 2.12. *Polytetrafluoroethylene structure.*

Source: https://en.wikipedia.org/wiki/Polytetrafluoroethylene.

PTFE has a waxy appearance and is a white solid. It is very flexible and tough, with an excellent electrical insulation characteristic. The coefficient of friction and surface energy are both very low, the former being the lowest of any solid. The combination of low coefficient of friction and low surface energy cause PTFE to have outstanding non-stick characteristics, a property which, a feature which causes numerous of the everyday applications of this polymer (Brown and Dattelbaum, 2005).

Moreover, due to its distinctive mechanical properties, PTFE has outstanding chemical resistance to a very wide variety of reagents, with the exemption of fluorine and molten alkali metals. There is no definite solvent for PTFE (Bauer et al., 1987). PTFE is an additional type of linear polymer, prepared by polymerization of the unsaturated monomer tetrafluoroethylene, $CF_2=CF_2$. Even though this structure should impart thermoplastic property to the polymer, PTFE does not display straight melting behavior. It does not seemingly liquify on heating, nor does it give a melt that will drift. As a substitute, it forms a translucent material of high-viscosity which breaks rather than moves when a suitable force is applied (Frantzides et al., 2002).

Because PTFE cannot be dissolved or melted, there are problems with fabricating articles out of the polymer. These are overcome by using a technique similar to powder metallurgy, in which granular particles are fused under high pressures and temperatures. The main domestic use for PTFE is on non-stick utensils such as frying pans. Industrially, the polymer is used for gaskets, pump parts, and laboratory equipment (Lappan et al., 2001).

2.6.5. Polyurethanes (PURs)

A functional group formed from the reaction of an isocyanate group with a hydroxy group is given the name 'Urethane' (Reaction 1.12).

$$\text{—N=C=O} + \text{HO—} \longrightarrow \text{—NHCOO—} \qquad (1.12)$$

Thermoset polymers, PURs, are formed from polyfunctional and di-isocyanates compounds comprising various hydroxy-groups. Normally the starting materials are themselves polymeric but include comparatively some monomer units in the molecule. Generally, low relative molar mass species of this type are called *oligomers.* Usual oligomers are polyethers and polyesters for the preparation of PURs. These are normally formed to include a little portion of monomeric trifunctional hydroxy compounds such as in the backbone, trimethylolpropane so that they comprise pendant hydroxyls which act as the sites of crosslinking (Kim et al., 1996; Petrović, 2008).

Flexible and rigid foams are the main use of PURs. There are several applications of rigid PURs, including insulating material for filling cavity walls of houses. In soft furnishing, flexible PURs have found many applications. They burn easily with a very toxic and smokey flame which is one of the considerable disadvantages of these PURs, and it has led to serious issues in accidental domestic fires. The applications of these materials for said purposes are nod diminishing (Gogolewski, 1989; Sardon et al., 2015).

2.6.6. Poly(Ether Ether Ketone)

Poly(ether ether ketone), is a crystalline high-temperature thermoplastic known as PEEK. 4,4-difluorobenzophenone is reacted with the anion of hydroquinone for its production by a step (condensation) process (Huang et al., 2001; Xing et al., 2004). A high boiling point polar solvent is used to perform this reaction, such as *N*-cyclohexyl-2-pyrrolidone (1.13).

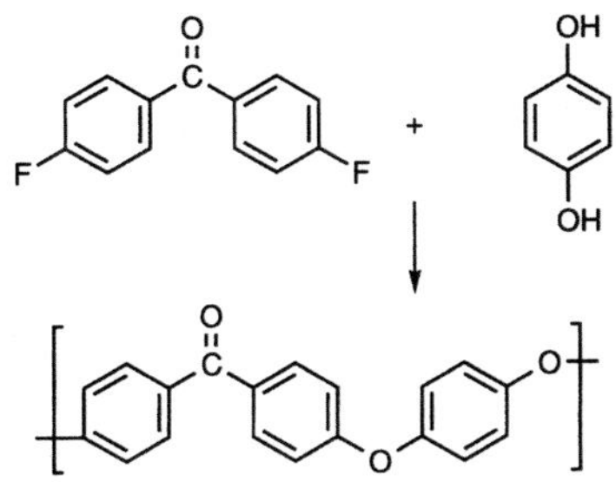

(1.13)

It can be used at temperatures of up to 250°C without damage to physical properties which indicates the thermal stability of PEEK. It has a melting point of 334°C (Sandler et al., 2002; Lakshmi et al., 2005). Copolymerization with monomers containing diphenyl units can reduce the degree of crystallinity, hence decreasing the degree of regularity in the linear array of linked phenyl rings. Industrial uses of PEEK take benefit of the excellent thermal properties and include application in high-performance aerospace and automotive, and in printed circuit boards (Jones et al., 1985; Cebe and Hong, 1986).

2.6.7. Silicones

Silicones are polymers with alternating oxygen and silicon atoms in the backbone in which the silicon atoms are also linked to organic groups (Schneider et al., 2008). Normal structures found are (1.4) and (1.5).

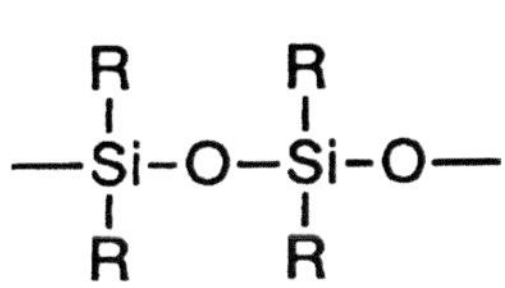

(1.4)

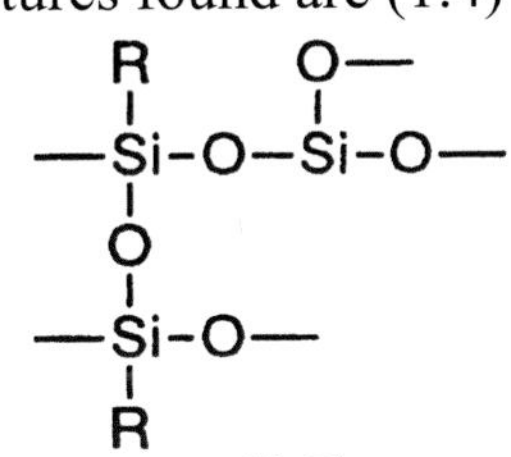

(1.5)

The likelihood of all types of morphology from low relative molar mass, linear molecules to three-dimensional systems means that silicones are present in a range of physical forms, from insoluble solids to fluids (Braley, 1970; Blundell, 1987; Colas, 2005).

2.7. NATURALLY OCCURRING POLYMERS

Various naturally present polymers have a technical application, including starch, cellulose, and its derivatives, starch, and rubber. Moreover, several significant biological materials, most particularly proteins, are made up of macromolecules. These will be discussed briefly in subsections (Wang et al., 1998; Shukla et al., 2012).

2.7.1. Cellulose

This is a very broadly accessible polymer since it is the key constituent of the cell walls of all plants. It is a carbohydrate of molecular formula $(C_6H_{10}O_5)_n$, where *n* runs to thousands. The cellulose molecules are built up from this substance effectively by removal and condensation of the elements of water and the cellulose 'monomer' is D-glucose (Figure 2.13) (Racles et al., 2016; Kottmann et al., 2017).

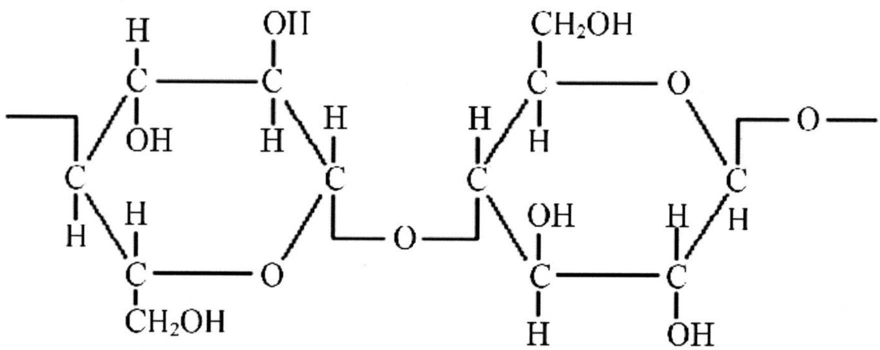

Cellulose

Figure 2.13. *The structural formula of cellulose.*

Source: https://www.researchgate.net/figure/Structure-of-cellulose_fig3_235407157.

Cellulose is not soluble in water but; glucose itself is highly soluble. This is principally a kinetic phenomenon; the hydroxy-groups in cellulose would readily form hydrogen bonds with water molecules, and therefore the cellulose macromolecule would take into aqueous solution (Williams and Craig, 1988; Owen, 2010). But these hydroxy groups cooperate with neighboring cellulose molecules, making it difficult for water molecules to solvate, or penetrate, the individual molecules. However, cellulose will dissolve in the rather strange medium aqueous ammoniacal cupric hydroxide, $Cu(NH_3)_4(OH)_2$ (Welch and Vandevivere, 1994; Davidson, 1996).

2.7.2. Starch

Starch is a broadly dispersed material that is present in seeds, roots, and fruits of plants. For industrial use, though potatoes and wheat are also used, corn is the main source. Starch is produced by grinding with water, filtering, centrifuging, and drying, a process that produces starch in granular form (Figure 2.14) (Umoren et al., 2008; Naseri-Nosar and Ziora, 2018).

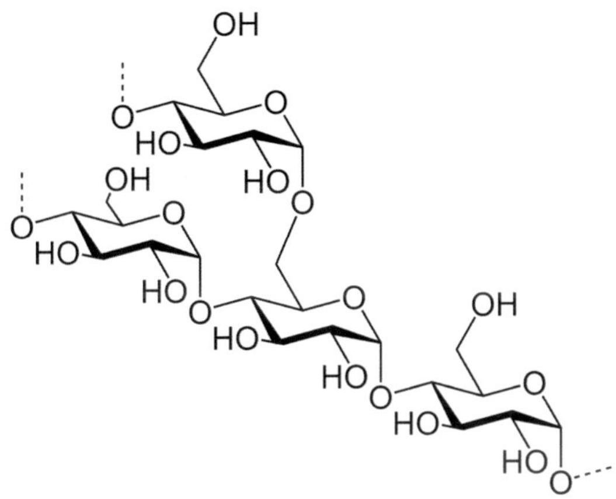

Figure 2.14. *The structural formula of starch.*

Source: https://en.wikipedia.org/wiki/Starch.

There are two components in starch, amylopectin and amylose. Varying sources of starch change the ratios of these components, with amylopectin typically prevailing and demonstrating some 70–85% of the total mass of the starch (Turbak et al., 1983; Jane, 1995; Fischer et al., 2008).

2.7.3. Natural Rubber

The juice of several tropical trees is the source of rubber, mainly the tree *Hevea brasiliensis.* Together with traces of fats, proteins, salts, and sterols, the juice is a latex comprising of dispersion of polymer phase at a concentration of about 35% by mass. The rubber is gained either by evaporation in the air or over flame coagulation of the latex with acid, either methanoic or ethanoic. The resultant of these processes is a cheese like substance, crumbly, and sometimes known as caoutchouc or raw rubber.

To establish mechanical properties that are the fundamental features of the rubber, i.e., called rubberlike elasticity, this raw rubber requires additional processing, and in specific lightly crosslinking. Vulcanization is the process used for this purpose (Maningat and Juliano, 1980).

In natural rubber, the polymer contains almost completely *cis*-poly(isoprene) (1.6). The molecules have relative molar mass usually ranging between 300,000 and 500,000 and are linear. In 1922, Staudinger developed the macromolecular nature of rubber when he hydrogenated the material attained a product that reserved its colloidal character, rather than producing pieces of low relative molar mass (Payne, 1962).

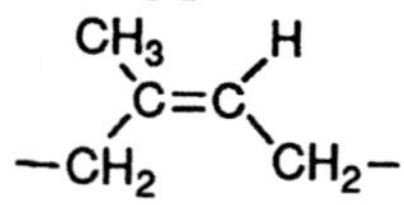

(1.6)

A process in which rubber molecules are lightly crosslinked to develop elasticity and reduce plasticity is known as vulcanization. Originally, it was used for the application of sulfur but is now used for any similar process of crosslinking. Though, sulfur remains the substance most abundantly used for this purpose.

2.7.4. Proteins

In biochemistry, the proteins are a group of macromolecular substances of pronounced significance. Their name gives evidence to this, in 1838, it was devised by Mulder from the Greek word 'proteins, meaning 'of first importance. They are present in all cells, both plant and animal, and take part in all cell functions. Formed from α-amino acids, proteins are linear PAs. An α-amino acid is one in which the amino group and carboxylic acid reside on the same carbon atom (1.7) (Figure 2.15).

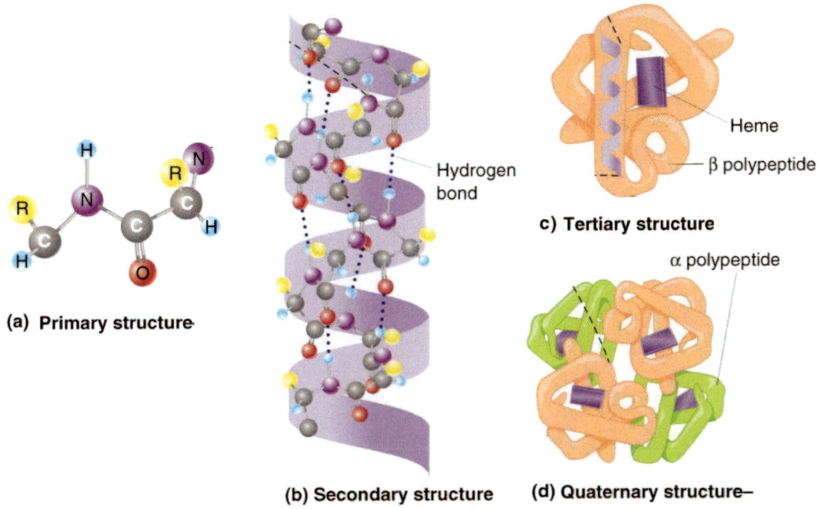

Figure 2.15. Structures of protein: (a) primary structure; (b) secondary structure; (c) tertiary structure; (d) quaternary structure.

Source: https://www.mun.ca/biology/scarr/iGen3_06-04.html.

There are 20 amino acids in nature, available for integration into the protein chain. They are organized in a characteristic and specific sequence along the molecule. This sequence is usually mentioned as the 'primary structure' of the protein. Also, the relative molar mass of the macromolecule is part of the primary structure.

$$\mathrm{R{-}\underset{\displaystyle NH_2}{\underset{|}{C}H}{-}COOH} \qquad (1.7)$$

The protein molecule adopts a distinctive arrangement, as a result of the specific sequence of the amino acid such as orderly folding or symmetrical coils. This is named the 'secondary structure.' Such folded structures and individuals' coils bring about a longer-range steady arrangement called the 'tertiary structure.' Finally, to form a complex, numerous protein molecules may link together this final grouping being called the 'quaternary structure.'

Several 100 amino acids in the primary structure are present in the large molecules of the proteins. During physiological processes, smaller units, called polypeptides may be prepared. There is no clear difference between a polypeptide and a protein; for a substance to be considered a protein, about 200 amino acid units in the primary structure is frequently taken to be the minimum. The variety in quaternary, tertiary, secondary, and primary structures of proteins means that regarding their chemical properties, few generalizations can be made. Some accomplish structural roles, such as the keratin (found in beaks and claws), the collagens (found in bone), and are insoluble in all solvents. Others, such as globulins and albumins of plasma, are very soluble in water. Still, others, which are partly lipophilic ('lipid-loving,' hence fat-soluble) and partly hydrophilic ('water-loving,' hence water-soluble), form part of membranes of cells (Bitinis et al., 2011).

2.7.5. Poly(3-Hydroxybutyrate)

Poly(3-hydroxybutyrate) (1.8) that behaves as an acceptable thermoplastic is a bacterial polyester, yet is biodegradable and can be formed from renewable agricultural feedstocks. It is usually not prepared in the pure state., but prepared alongside minor quantities of poly(3-hydroxy valerate). The ratio of propionic acid and glucose in the medium in which bacteria are present and perform their metabolic processes determines the ratio of these two polymers in a given sample.

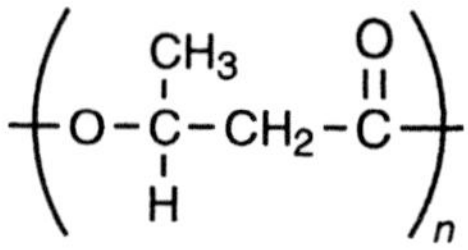

(1.8)

The carbon atom which conveys the methyl group is chiral, but biosynthesis is stereoselective and produces a natural polymer with the *R* configuration. The polymer melts at about 80°C and is partly crystalline thermoplastic.

REFERENCES

1. Abir-Am, P. G., (1992). The politics of macromolecules: Molecular biologists, biochemists, and rhetoric. *Osiris, 7*, 164–191.
2. Ali, U., Karim, K. J. B. A., & Buang, N. A., (2015). A review of the properties and applications of poly(methyl methacrylate) (PMMA). *Polymer Reviews, 55*(4), 678–705.
3. Anzano, J., Lasheras, R. J., Bonilla, B., & Casas, J., (2008). Classification of polymers by determining of C1: C2: CN: H: N: O ratios by laser-induced plasma spectroscopy (LIPS). *Polymer Testing, 27*(6), 705–710.
4. Aranda, P., & Ruiz-Hitzky, E., (1992). Poly(ethylene oxide)-silicate intercalation materials. *Chemistry of Materials, 4*(6), 1395–1403.
5. Arturson, G., (1979). Microvascular permeability to macromolecules in thermal injury. *Acta Physiologica Scandinavica. Supplementum, 463*, 111–122.
6. Bauer, J. J., Salky, B. A., Gelernt, I. M., & Kreel, I., (1987). Repair of large abdominal wall defects with expanded polytetrafluoroethylene (PTFE). *Annals of Surgery, 206*(6), 765.
7. Bitinis, N., Verdejo, R., Cassagnau, P., & Lopez-Manchado, M. A., (2011). Structure and properties of polylactide/natural rubber blends. *Materials Chemistry and Physics, 129*(3), 823–831.
8. Bloembergen, N., (1996). Nonlinear optics of polymers: Fundamentals and applications. *Journal of Nonlinear Optical Physics and Materials, 5*(1), 1–8.
9. Blundell, D. J., (1987). On the interpretation of multiple melting peaks in poly(ether ether ketone). *Polymer, 28*(13), 2248–2251.
10. Boeykens, S., & Vázquez, C., (2004). Preliminary classification of polymers by using total-reflection X-ray fluorescence spectra. *Spectrochimica Acta Part B: Atomic Spectroscopy, 59*(8), 1189–1192.
11. Braley, S., (1970). The chemistry and properties of the medical-grade silicones. *Journal of Macromolecular Science-Chemistry, 4*(3), 529–544.
12. Braun, D., (2004). Poly(vinyl chloride) on the way from the 19th century to the 21st century. *Journal of Polymer Science Part A: Polymer Chemistry, 42*(3), 578–586.

13. Brown, E. N., & Dattelbaum, D. M., (2005). The role of crystalline phase on fracture and microstructure evolution of polytetrafluoroethylene (PTFE). *Polymer, 46*(9), 3056–3068.
14. Bucknall, C. B., & Gilbert, A. H., (1989). Toughening tetrafunctional epoxy resins using polyetherimide. *Polymer, 30*(2), 213–217.
15. Busico, V., & Cipullo, R., (2001). Microstructure of polypropylene. *Progress in Polymer Science, 26*(3), 443–533.
16. Cebe, P., & Hong, S. D., (1986). Crystallization behavior of poly(ether-ether-ketone). *Polymer, 27*(8), 1183–1192.
17. Chandrasekhar, P., (1999). *Conducting Polymers, Fundamentals and Applications, 1*, 10–23.
18. Colas, A., (2005). Silicones: Preparation, properties, and performance. *Dow Corning, Life Sciences, 1*, 1–22.
19. Cornelissen, J. J., Fischer, M., Sommerdijk, N. A., & Nolte, R. J., (1998). Helical superstructures from charged poly(styrene)-poly(isocyanodipeptide) block copolymers. *Science, 280*(5368), 1427–1430.
20. Costa, L., Di Montelera, L. R., Camino, G., Weil, E. D., & Pearce, E. M., (1997). Structure-charring relationship in phenol-formaldehyde type resins. *Polymer Degradation and Stability, 56*(1), 23–35.
21. Covas, J. A., Agassant, J. F., Diogo, A. C., Vlachopoulos, J., & Walters, K., (2013). *Rheological Fundamentals of Polymer Processing* (Vol. 302, pp. 1–29). Springer Science & Business Media.
22. Davidson, R. S., (1996). The photodegradation of some naturally occurring polymers. *Journal of Photochemistry and Photobiology B: Biology, 33*(1), 3–25.
23. Dhote, V. K., Dhote, K., Pandey, S. P., Shukla, T., Maheshwari, R., Mishra, D. K., & Tekade, R. K., (2019). Fundamentals of polymers science applied in pharmaceutical product development. In: *Basic Fundamentals of Drug Delivery* (Vol. 1, pp. 85–112). Academic Press.
24. Doi, M., & Edwards, S. F., (1978). Dynamics of rod-like macromolecules in concentrated solution; Part 2. *Journal of the Chemical Society, Faraday Transactions 2: Molecular and Chemical Physics, 74*, 918–932.
25. El Mansouri, N. E., Yuan, Q., & Huang, F., (2011). Characterization of alkaline lignins for use in phenol-formaldehyde and epoxy resins. *BioResources, 6*(3), 2647–2662.

26. El-Magied, M. O. A., Elshehy, E. A., Manaa, E. S. A., Tolba, A. A., & Atia, A. A., (2016). Kinetics and thermodynamics studies on the recovery of thorium ions using amino resins with magnetic properties. *Industrial & Engineering Chemistry Research, 55*(43), 11338–11345.
27. Fischer, S., Thümmler, K., Volkert, B., Hettrich, K., Schmidt, I., & Fischer, K., (2008). Properties and applications of cellulose acetate. In: *Macromolecular symposia* (Vol. 262, No. 1, pp. 89–96). Weinheim: Wiley-VCH Verlag.
28. Frantzides, C. T., Madan, A. K., Carlson, M. A., & Stavropoulos, G. P., (2002). A prospective, randomized trial of laparoscopic polytetrafluoroethylene (PTFE) patch repair vs simple cruroplasty for large hiatal hernia. *Archives of Surgery, 137*(6), 649–652.
29. Gao, M., Zhu, K., Sun, Y. J., & Sun, C., (2004). Thermal degradation of wood treated with amino resins and amino resins modified with phosphate in nitrogen. *Journal of Fire Sciences, 22*(6), 505–515.
30. Garbarczyk, J., & Paukszta, D., (1985). Influence of additives on the structure and properties of polymers. *Colloid and Polymer Science, 263*(12), 985–990.
31. Gogolewski, S., (1989). Selected topics in biomedical polyurethanes: A review. *Colloid and Polymer Science, 267*(9), 757–785.
32. Helfand, E., (1975). Theory of inhomogeneous polymers: Fundamentals of the gaussian random-walk model. *The Journal of Chemical Physics, 62*(3), 999–1005.
33. Huang, R. Y. M., Shao, P., Burns, C. M., & Feng, X., (2001). Sulfonation of poly(ether ether ketone) (PEEK): Kinetic study and characterization. *Journal of Applied Polymer Science, 82*(11), 2651–2660.
34. Israelachvili, J., (1997). The different faces of poly(ethylene glycol). *Proceedings of the National Academy of Sciences, 94*(16), 8378–8379.
35. Jane, J. J. M. S., (1995). Starch properties, modifications, and applications. *Journal of Macromolecular Science, Part A: Pure and Applied Chemistry, 32*(4), 751–757.
36. Jia, Z., Wang, Z., Xu, C., Liang, J., Wei, B., Wu, D., & Zhu, S., (1999). Study on poly(methyl methacrylate)/carbon nanotube composites. *Materials Science and Engineering: A, 271*(1, 2), 395–400.
37. Jin, J., Rafiq, R., Gill, Y. Q., & Song, M., (2013). Preparation and characterization of high performance of graphene/nylon nanocomposites. *European Polymer Journal, 49*(9), 2617–2626.

38. Jones, A. T., Aizlewood, J. M., & Beckett, D. R., (1964). Crystalline forms of isotactic polypropylene. *Die Makromolekulare Chemie: Macromolecular Chemistry and Physics, 75*(1), 134–158.
39. Jones, D. P., Leach, D. C., & Moore, D. R., (1985). Mechanical properties of poly(ether-ether-ketone) for engineering applications. *Polymer, 26*(9), 1385–1393.
40. Kang, S. C., & Chung, D. W., (2000). The synthesis and frictional properties of lubricant-impregnated cast nylons. *Wear, 239*(2), 244–250.
41. Karplus, M., (2003). Molecular dynamics of biological macromolecules: A brief history and perspective. *Biopolymers: Original Research on Biomolecules, 68*(3), 350–358.
42. Kim, B. K., Lee, S. Y., & Xu, M., (1996). Polyurethanes having shape memory effects. *Polymer, 37*(26), 5781–5793.
43. Kottmann, A., Mejía, E., Hémery, T., Klein, J., & Kragl, U., (2017). Recent developments in the preparation of silicones with antimicrobial properties. *Chemistry: An Asian Journal, 12*(11), 1168–1179.
44. Kulcke, A., Gurschler, C., Spöck, G., Leitner, R., & Kraft, M., (2003). On-line classification of synthetic polymers using near-infrared spectral imaging. *Journal of Near Infrared Spectroscopy, 11*(1), 71–81.
45. Lakshmi, R. M., Choudhary, V., & Varma, I. K., (2005). Sulphonated poly(ether ether ketone): Synthesis and characterization. *Journal of Materials Science, 40*(3), 629–636.
46. Lappan, U., Geißler, U., Häußler, L., Jehnichen, D., Pompe, G., & Lunkwitz, K., (2001). Radiation-induced branching and crosslinking of poly(tetrafluoroethylene) (PTFE). *Nuclear Instruments and Methods in Physics Research Section B: Beam Interactions with Materials and Atoms, 185*(1–4), 178–183.
47. Leitner, R., Mairer, H., & Kercek, A., (2003). Real-time classification of polymers with NIR spectral imaging and blob analysis. *Real-Time Imaging, 9*(4), 245–251.
48. Maningat, C. C., & Juliano, B. O., (1980). Starch lipids and their effect on rice starch properties. *Starch-Stärke, 32*(3), 76–82.
49. Migliaresi, C., & Pegoretti, A., (2002). Fundamentals of polymeric composite materials. In: *Integrated Biomaterials Science* (Vol 1., pp. 69–117). Springer, Boston, MA.

50. Minale, M., Moldenaers, P., & Mewis, J., (1997). Effect of shear history on the morphology of immiscible polymer blends. *Macromolecules, 30*(18), 5470–5475.
51. Morgan, A. B., & Gilman, J. W., (2013). An overview of flame retardancy of polymeric materials: Application, technology, and future directions. *Fire and Materials, 37*(4), 259–279.
52. Moubarik, A., Pizzi, A., Allal, A., Charrier, F., & Charrier, B., (2009). Cornstarch and tannin in phenol-formaldehyde resins for plywood production. *Industrial Crops and Products, 30*(2), 188–193.
53. Muñoz-Bonilla, A., & Fernández-García, M., (2012). Polymeric materials with antimicrobial activity. *Progress in Polymer Science, 37*(2), 281–339.
54. Naseri-Nosar, M., & Ziora, Z. M., (2018). Wound dressings from naturally-occurring polymers: A review on homopolysaccharide-based composites. *Carbohydrate Polymers, 189*, 379–398.
55. Nguon, O., Lagugné-Labarthet, F., Brandys, F. A., Li, J., & Gillies, E. R., (2018). Microencapsulation by in situ polymerization of amino resins. *Polymer Reviews, 58*(2), 326–375.
56. Novikov, V. U., & Kozlov, G. V., (2000). Structure and properties of polymers in terms of the fractal approach. *Russian Chemical Reviews, 69*(6), 523–549.
57. Nucci, M. L., Shorr, R., & Abuchowski, A., (1991). The therapeutic value of poly(ethylene glycol)-modified proteins. *Advanced Drug Delivery Reviews, 6*(2), 133–151.
58. Owen, M. J., (2010). Properties and applications of silicones. In: *Advances in Silicones and Silicone-Modified Materials* (pp. 13–18). American Chemical Society.
59. Padden, Jr. F. J., & Keith, H. D., (1959). Spherulitic crystallization in polypropylene. *Journal of Applied Physics, 30*(10), 1479–1484.
60. Papkov, S. P., (1982). Classification of polymers according to their phase state. *Polymer Science USSR*, *24*(8), 1939–1944.
61. Payne, A. R., (1962). The dynamic properties of carbon black-loaded natural rubber vulcanizates; Part I. *Journal of Applied Polymer Science, 6*(19), 57–63.
62. Petrović, Z. S., (2008). Polyurethanes from vegetable oils. *Polymer Reviews, 48*(1), 109–155.

63. Poljansek, I., & Krajnc, M., (2005). Characterization of phenol-formaldehyde prepolymer resins by in line FT-IR spectroscopy. *Acta Chimica Slovenica, 52*(3), 238.

64. Prakash, J., Pivin, J. C., & Swart, H. C., (2015). Noble metal nanoparticles embedding into polymeric materials: From fundamentals to applications. *Advances in Colloid and Interface Science, 226*, 187–202.

65. Prucker, O., & Rühe, J., (1998). Synthesis of poly(styrene) monolayers attached to high surface area silica gels through self-assembled monolayers of azo initiators. *Macromolecules, 31*(3), 592–601.

66. Racles, C., Cozan, V., Bele, A., & Dascalu, M., (2016). Polar silicones: Structure-dielectric properties relationship. *Designed Monomers and Polymers, 19*(6), 496–507.

67. Rae, P. J., & Dattelbaum, D. M., (2004). The properties of poly(tetrafluoroethylene) (PTFE) in compression. *Polymer, 45*(22), 7615–7625.

68. Ringsdorf, H., (1975). Structure and properties of pharmacologically active polymers. In: *Journal of Polymer Science: Polymer Symposia* (Vol. 51, No. 1, pp. 135–153). New York: Wiley Subscription Services, Inc., A Wiley Company.

69. Roth, C. B., & Baglay, R. R., (2016). Fundamentals of polymers and glasses. In: *Polymer Glasses* (Vol. 1, pp. 19–38). CRC Press.

70. Russell, T. P., Hjelm, Jr. R. P., & Seeger, P. A., (1990). Temperature dependence of the interaction parameter of polystyrene and poly(methyl methacrylate). *Macromolecules, 23*(3), 890–893.

71. Sandler, J., Werner, P., Shaffer, M. S., Demchuk, V., Altstädt, V., & Windle, A. H., (2002). Carbon-nanofiber-reinforced poly(ether ether ketone) composites. *Composites Part A: Applied Science and Manufacturing, 33*(8), 1033–1039.

72. Sardon, H., Pascual, A., Mecerreyes, D., Taton, D., Cramail, H., & Hedrick, J. L., (2015). Synthesis of polyurethanes using organocatalysis: A perspective. *Macromolecules, 48*(10), 3153–3165.

73. Schneider, F., Fellner, T., Wilde, J., & Wallrabe, U., (2008). Mechanical properties of silicones for MEMS. *Journal of Micromechanics and Microengineering, 18*(6), 065008.

74. Schurer, J. W., De Boer, A., & Challa, G., (1975). Influence of tacticity of poly(methyl methacrylate) on the compatibility with poly(vinyl chloride). *Polymer, 16*(3), 201–204.

75. Seymour, L. W., (1992). Passive tumor targeting of soluble macromolecules and drug conjugates. *Critical Reviews in Therapeutic Drug Carrier Systems, 9*(2), 135–187.

76. Shukla, D., Negi, Y. S., Uppadhyaya, J. S., & Kumar, V., (2012). Synthesis and modification of poly(ether ether ketone) and their properties: A review. *Polymer Reviews, 52*(2), 189–228.

77. Tobolsky, A., & Eyring, H., (1943). Mechanical properties of polymeric materials. *The Journal of Chemical Physics, 11*(3), 125–134.

78. Torlakoğlu, A., & Güçlü, G., (2009). Alkyd-amino resins based on waste PET for coating applications. *Waste Management, 29*(1), 350–354.

79. Turbak, A. F., Snyder, F. W., & Sandberg, K. R., (1983). Micro fibrillated cellulose, a new cellulose product: properties, uses, and commercial potential. In: *J. Appl. Polym. Sci. Appl. Polym. Symp.* (Vol. 37, No. 9, pp. 815–827).

80. Umoren, S. A., Ogbobe, O., Igwe, I. O., & Ebenso, E. E., (2008). Inhibition of mild steel corrosion in acidic medium using synthetic and naturally occurring polymers and synergistic halide additives. *Corrosion Science, 50*(7), 1998–2006.

81. Wang, F., Chen, T., & Xu, J., (1998). Sodium sulfonate-functionalized poly(ether ether ketone) s. *Macromolecular Chemistry and Physics, 199*(7), 1421–1426.

82. Welch, S. A., & Vandevivere, P., (1994). Effect of microbial and other naturally occurring polymers on mineral dissolution. *Geomicrobiology Journal, 12*(4), 227–238.

83. Wilhelm, M., Zhao, C. L., Wang, Y., Xu, R., Winnik, M. A., Mura, J. L., & Croucher, M. D., (1991). Poly(styrene-ethylene oxide) block copolymer micelle formation in water: A fluorescence probe study. *Macromolecules, 24*(5), 1033–1040.

84. Williams, D. J., (1984). Organic polymeric and non-polymeric materials with large optical nonlinearities. *Angewandte Chemie International Edition in English, 23*(9), 690–703.

85. Williams, J. R., & Craig, R. G., (1988). Physical properties of addition silicones as a function of composition. *Journal of Oral Rehabilitation, 15*(6), 639–650.

86. Wohrle, D., & Pomogailo, A. D., (2003). *Metal Complexes and Metals in Macromolecules: Synthesis, Structure and Properties* (Vol. 1, pp. 1–22) Wiley, Weinheim.

87. Wu, D. Y., Meure, S., & Solomon, D., (2008). Self-healing polymeric materials: A review of recent developments. *Progress in Polymer Science, 33*(5), 479–522.

88. Wu, S., (1970). Surface and interfacial tensions of polymer melts; II. Poly(methyl methacrylate), poly(n-butyl methacrylate), and polystyrene. *The Journal of Physical Chemistry, 74*(3), 632–638.

89. Xing, P., Robertson, G. P., Guiver, M. D., Mikhailenko, S. D., Wang, K., & Kaliaguine, S., (2004). Synthesis and characterization of sulfonated poly(ether ether ketone) for proton exchange membranes. *Journal of Membrane Science, 229*(1, 2), 95–106.

90. Xu, J., & Xu, Z. L., (2002). Poly(vinyl chloride) (PVC) hollow fiber ultrafiltration membranes prepared from PVC/additives/solvent. *Journal of Membrane Science, 208*(1, 2), 203–212.

91. Yamini, S., & Young, R. J., (1980). The mechanical properties of epoxy resins. *Journal of Materials Science, 15*(7), 1823–1831.

92. Zachariades, A. E., & Porter, R. S., (1979). Reversible plasticization of nylons 6 and 11 with anhydrous ammonia and their deformation by solid-state coextrusion. *Journal of Applied Polymer Science, 24*(5), 1371–1382.

Chapter 3

Introduction of Fiber-Reinforced Polymers

CONTENTS

3.1. Introduction 82
3.2. Basic Concepts and Manufacturing Process 86
3.3. Types of Polymer Composites 88
3.4. Fiber Manufacturing Process 93
3.5. Composite Molding 96
3.6. Resin Infusion 101
3.7. Glass Fiber Material 103
3.8. Aramid Fiber Material Process 104
3.9. FRP Applications 104
3.10. Mechanical Properties Measurements 107
3.11. Real Polymer's Mechanical Properties 112
References 113

3.1. INTRODUCTION

Fiber-reinforced plastic (FRP) is a composite material that is made by reinforcing polymer matrix with fibers. *FRP* is also termed as Fiber-reinforced polymers. Usually, fibers such as aramid, carbon, or glass are used. Sometimes other fibers, for example, asbestos, wood, or paper, are also used. Usually, the polymer is one of the phenol-formaldehyde resins, polyester or vinyl ester (VE) thermosetting plastic, and epoxy. Commonly, FRPs are used in the construction, automotive, aerospace, and marine industries (Fu and Lauke, 1996; Weissenböck et al., 2014).

Composite materials are naturally occurring or engineered materials that are made up of at least two constituent materials, which have significantly different chemical or physical properties. These properties remain distinct and separate within the finished structure. The composites usually have stiff and strong fibers in a matrix that is less stiff and weaker. The aim is to make a stiff and strong component, often having a low density (Wegian and Abdalla, 2005). Commonly, commercial material contains carbon or glass fibers in thermosetting polymers-based matrices, such as polyester or epoxy resins. Sometimes, thermoplastic polymers are preferred, as, after initial production, they are moldable. In further classes of composite, the matrix is ceramic or metal. Still, most parts of these composites are in a developmental stage. Many crucial problems like high manufacturing costs are still not figured out (Tong et al., 2002; Hinton et al., 2004). Moreover, the reasons for adding the fibers in these composites are often complex; e.g., improvements might be sought in wear, thermal stability, wear, fracture toughness, etc. (Figure 3.1).

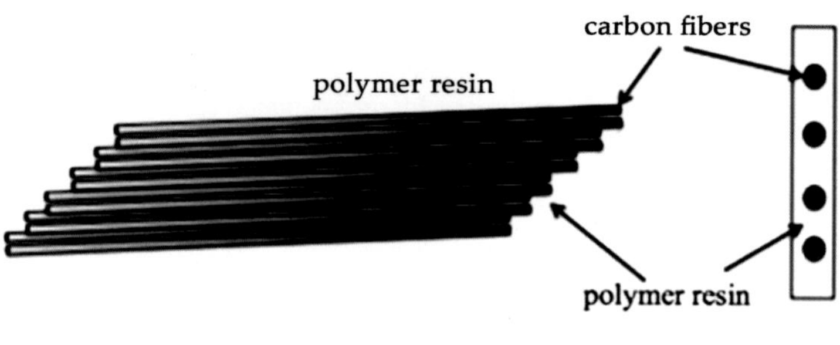

Figure 3.1. Fiber-reinforced polymer.

Source: https://www.researchgate.net/figure/A-typical-structure-of-carbon-fiber-reinforced-polymer-CFRP_fig1_333705919.

In all types of advanced engineering structures, FRP composites are being used. They are widely used in spacecraft, helicopters, aircraft, ships, boats, offshore platforms, sports goods, civil infrastructures such as buildings and bridges, automobiles, and chemical processing equipment. The rate at which this usage is growing is really impressive (González and LLorca, 2007; Landesmann et al., 2015). They are being not only used in their existing markets but also spreading their domain in new markets such as civil structures and biomedical devices. Over the recent years, the development of advanced FRP materials has served as an important factor in increasing the applications of composites. This includes the advancement in resin systems (better performance) and new reinforcement styles, such as nanoparticles and carbon nanotubes. This book gives a recent account of the mechanical properties, applications of 3D FRP composites, fabrication, impact tolerance, and delamination resistance (Jain and Lee, 2012; Hofstätter et al., 2017).

FRP composites are being considered as a substitute and/or an enhancement to infrastructure systems or components, which are traditionally constructed through civil engineering materials, namely steel and concrete. FRP composites exhibit high specific stiffness and strength, are non-corrosive, and lightweight. They can be easily constructed and tailored for satisfying performance requirements. Because of these beneficial characteristics, they are being included in rehabilitation and new constructions of structures. They can be introduced as reinforcement in modular structures, concrete, formwork, bridge decks, and external reinforcement for seismic upgrade and strengthening (Boztepe et al., 2017; Aamir et al., 2019).

In numerous professional organizations and research laboratories around the world, the applicability of FRP reinforcements to concrete structures has been actively studied, and it is found that it can be used as a substitute for prestressing tendons or steel bars. These reinforcements provide numerous advantages such as high tensile strength, corrosion resistance, ease of handling, non-magnetic properties, and lightweight. However, they usually give a linear elastic response in a brittle failure (explained as tension up to failure) and a relatively poor shear or transverse resistance. Similarly, when exposed to fire or high temperatures, they offer poor resistance. Upon bending, significant strength is lost by them and they show high sensitiveness to stress-rupture effects. Furthermore, in contrast to conventional prestressing tendons or steel reinforcing bars, FRPs are costly (whether considered on the basis of per unit weight or force-carrying capacity). From the viewpoint of structural engineering, the major issues with FRP reinforcements are that they offer extremely low shear strength and lack of plastic behavior in the transverse direction (Rana and Purohit, 2017). These kinds of characteristics can result in premature tendon rupture, especially when combined effects, like the one in reinforced concrete beams at shear-cracking planes where dowel action exists, are present. The dowel action reduces shear resistance and residual tensile in the tendon. Limitations of use and solutions have been recommended, and still, continuous improvements are being offered. It is expected that the increased market demand and share will significantly decrease the unit cost of FRP reinforcements. However, even today, a lot of FRP reinforcements applications are justifiable and cost-effective. Such examples include the bonded FRP plates or sheets usage in strengthening and repairing of concrete structures and the FRP fabrics or textiles or meshes usage in thin cement products. In relative terms, the cost of rehabilitation and repair of a structure is substantially higher than the initial structure's cost. Generally, repairing requires a high commitment in labor for a small volume of repair materials. Furthermore, the material's cost becomes secondary in developed countries, as the cost of labor is so high. Thus, better durability and performance of the repair material make the repair more cost-effective. Hence, material cost is not a problem in repair, and it implies the fact that the high cost of FRP repair materials is not a constraining drawback (Figure 3.2) (Boztepe et al., 2017).

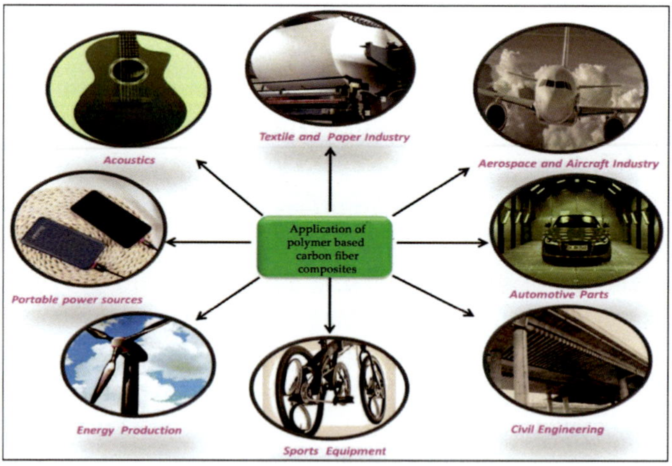

Figure 3.2. Applications in which carbon fiber reinforced composites are used.

Source: https://link.springer.com/article/10.1007/s42114-018-0072-z.

In a sustainable environment, the argument for FRP composites is questionable when considering only material and energy resources. However, this conclusion has to be evaluated in the lights of advantages that FRP composites offer in regards to considerations such as:

- Lighter weight;
- Longer lasting;
- Rehabilitating existing structures;
- Space systems;
- Seismic upgrades;
- Ocean environments;
- Higher-strength;
- Higher performance;
- Defense systems.

Environmental concerns, in FRP composite's case, appear as a barrier against its feasibility as a sustainable material, particularly when considering air pollution, acidification, smog, and fossil fuel depletion associated with its production. Moreover, there exists a limited ability for recycling FRP composites and, unlike timber and steel, we cannot use their structural components for performing a similar function in some other structures. However, the FRP composite's environmental impact evaluation in infrastructure applications (especially by the means of life cycle analysis) reveals indirect and direct advantages that, in contrast to conventional materials, are more competitive.

Since the introduction of composite materials, they have developed greatly. However, many requirements need to be fulfilled before using composite materials as a substitute for conventional materials to maintain a sustainable environment:

- Standardized durable characterization data availability for these materials;
- Integration of methods and durability data of structural members (that utilize FRP composites) service life prediction;
- Development of techniques and methods for materials selection on the basis of life cycle assessments of systems and structural components.

Composites, in order to be regarded as a viable alternative, have to be economically and structurally feasible. Many studies about the composite material's structural feasibility are present in the literature (Erki and Rizkalla, 1993). However, from the point of view of a life cycle approach, very few studies are available on the environmental and economic feasibility of these materials. This is because of the reason that only economic costs are taken into the comparison and lack of data, as only short-term data is available. Moreover, there is a need of determining the long-term effects on these materials. There exists a need to assess the potential to recycle these materials, the constituent material's sustainability, and byproducts of the production. This has to be assessed for determining the fact that whether these composite materials can serve a part in a sustainable environment or not. Hence, in this chapter, we will focus on the physicochemical properties of composites and polymers used in Civil Engineering.

3.2. BASIC CONCEPTS AND MANUFACTURING PROCESS

Generally, polymerization, polycondensation, or polyaddition are used to manufacture synthetic polymers. The polymers are combined with different agents for enhancing or altering the material properties of polymers. This results in plastic. The Composite plastics might be a heterogeneous or homogeneous mix (Zerbst et al., 2021). Composite plastics are formed when at least two homogeneous materials, having different material properties, bond together for deriving a product that has certain desired mechanical properties (and material). The FRPs (also known as fiber-reinforced polymers) fall under the class of composite plastics that only uses fiber materials, instead of the one mixed with a polymer, for mechanically enhancing the elasticity and strength of plastics. Without fiber reinforcement, the original plastic material has termed the matrix. It is a tough plastic (though it is relatively weak) that stronger stiffer reinforcing fibers or filaments reinforce. In an FRP, the extent that elasticity and strength are enhanced depends on the mechanical properties of the matrix and fiber, their volume, and the orientation and length of the fiber within the matrix. By definition, matrix reinforcement occurs when increased elasticity or strength, relative to the elasticity and strength of the matrix alone, is exhibited by the FRP material (Figure 3.3) (Schiling et al., 1999; Babuska and Oden, 2004).

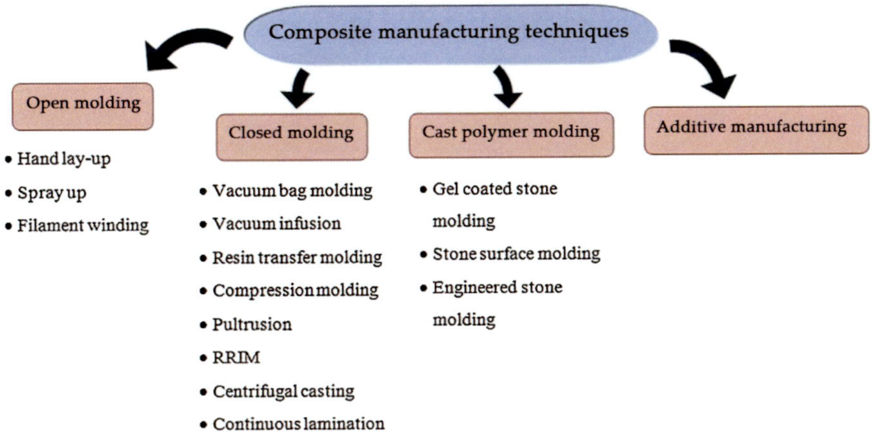

Figure 3.3. Different manufacturing processes of composite.

Source: https://www.sciencedirect.com/science/article/pii/S2238785419312086.

In contrast to other construction materials such as metals and ceramics, polymers are different as they have macromolecular nature. Their long-chain structure, which is covalently bonded, characterizes them as macromolecules and determines their processability (through their averaged molecular weight (Mn)) like deep draw-, spin-, blow-, generally melt-formability. The mechanical strength is determined by the number averaged molecular weight (Borsato, 2017). High molecular weights provide advantages for properties like wear, impact resistance, strain-to-break, etc. Hence, natural limits are met, as too high elongational and shear viscosities are yielded by too high molecular weights. This makes polymers unprocessable. UHMWPE's, ultrahigh-molecular-weight-poly-ethylenes, and PTFE's, poly-tetra-fluoroethylene, are prime examples. Also, the garbage bags and high-performance fibers, used for bullet-proof vests are made of polyethylene (PE). These high-performance fibers resulting in mechanical properties, having moduli of 150 GPa and can be strengthened to 4 GPa, gives the potential optimal use of the structure of polymers yields, linked with their low density (Miller and Tang, 2009). If we think about polymers, it becomes obvious that why the polymeric concept is used by living nature for building its structures, instead of solely depending on high strength applications like spider webs, wood, or silk (Meijer and Govaert, 2005).

3.3. TYPES OF POLYMER COMPOSITES

Any material that is made of at least two components is known as composite. A lot of composites are present around us, i.e., concrete. It is made of sand, cement, gravel, and often steel rods to reinforce it. The shiny balloons are made of a composite. It contains an aluminum foil sheet and a polyester sheet. The polymer composites are comprised of polymers or polymers along with various types of materials (Meijer and Govaert, 2005). The fiber-reinforced composites are the ones among which fiber made from one material is embedded in a different material (Zhang et al., 2007; Thakur et al., 2014).

3.3.1. Synthetic Polymer Composites

The polymer composites are the compositions or combinations comprising at least two materials as separate phases, among which at least one material has to be a polymer. Levels of properties or unique combinations can be obtained by combining a polymer and other material, e.g., another polymer, carbon, or glass. Synthetic polymeric composites examples include thermosetting or thermoplastic resins, polymer-, carbon-, glass-fiber-reinforced, polymer blends, mica- or silica-reinforced resins, polymer-impregnated or polymer-bonded wood or concrete, and carbon-reinforced rubber (Shi, 2009; Jain and Lee, 2012). It is useful to consider materials such as crystalline polymers and coatings as composites. Crystallites in a polymer matrix are termed crystalline polymers. Bone and wood are naturally occurring composites. Bone contains minerals bonded with collagen, while wood contains cellulosic fibers bonded with lignin. In contrast to this, polymeric compositions compounded with low proportions of processing aids or pigments or with a plasticizer are not generally considered composites (Hnat, 1991; Tooley, 1991).

The aim is to improve stiffness, strength, or dimensional stability, or toughness by embedding fibers or particles in a binding phase or matrix. Another aim is the use of readily available and inexpensive fillers for extending a scarce or more expensive resin. This aim is important as petroleum supplies turn out to be less reliable and costlier. Filler (i.e., glass spheres) are used by other applications for improving processability, incorporating the dry-lubricant particles, i.e., molybdenum sulfide, for making a self-lubricating bearing. Filler is also used for reducing permeability (Lu et al., 1998; Webb, 2009). Commonly used fiber-reinforced polymer composites are usually based on the mat, cloth, roving, or glass fibers embedded in a matrix of polyester or epoxy resin. A high level of stiffness and strength are conferred by reinforced thermosetting resins that contain polyaramids, boron, and especially carbon fibers. Carbon-fiber composites are relatively five times stiffer than the ones of steel. Many applications, due to these excellent properties, are uniquely suited for polyester and epoxy composites, such as parts for automobiles, chemical reaction vessels, components in new jet aircraft, rocket motor cases, and boat hulls (Figure 3.4) (Masuelli, 2013; Al-Mutuki et al., 2016).

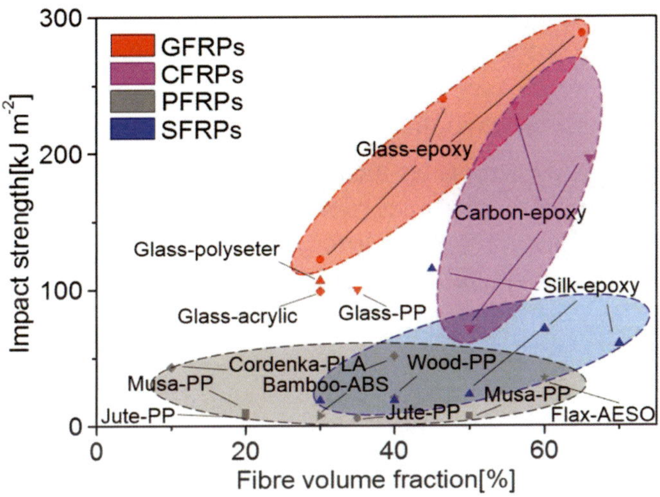

Figure 3.4. The impact strength comparison of polymer composites (at room temperature) reinforced by silk fibers, glass fibers, plant fibers, and carbon fibers.

Source: https://www.nature.com/articles/s41598-017-11919-1.

Reinforced thermosetting resins have the most dramatic properties, i.e., polyester, and epoxy resins, many reinforced thermoplastic resins can give significant improvements. The resins that are available as glass-reinforced composition include polyesters, PE, and polycarbonates (PC). By injection molding, the combination of one-step, inexpensive fabrication with improved properties has helped reinforced thermoplastics in replacing metals in numerous applications in automobiles, instruments, tools, and appliances (Auborg et al., 1991; Bauhofer and Kovacs, 2009).

Different matrices are possible for developing other composite systems: e.g., polyimide (PI) resins give a high-performance composite and are outstanding matrices for glass fibers. Different fibers, such as various metallic fibers, single-crystal ceramic whiskers (like sapphire), and polymers (like poly(vinyl alcohol), are of potential interest (Narula et al., 1996; Ohama, 1997).

Natural rubber latex, polyisoprene, was used by people living in Central and South America to make things like boots, rubber balls, and gloves. They took cotton fabric in two layers and embedded them in polyisoprene (natural rubber), making a three-layered sandwich (remember, cotton is made by cellulose, which is a natural polymer). This was used for making good raincoats, as cotton makes it comfortable for wearing and the rubber makes it waterproof. The properties of both, cotton, and rubber, are present in this material. Similarly, we have combined the comfort of cotton and the water-resistance of polyisoprene in this case (Chand, 2000; Wang et al., 2003).

Usually, modern composites are made of matrix and fiber. Mostly, the used fiber is glass, though sometimes carbon fiber, Kevlar, or PE is also used. The matrix is most often a thermoset like a PI, polydicyclopentadiene, or an epoxy resin. For making the matrix stronger, the fiber is embedded in it. Fiber-reinforced composites are light and strong. They weigh much less than steel but are often stronger than them. Hence, composites can be used for the making of automobiles lighter and will be more fuel-efficient (Suzuki, 1994; Gullapalli and Wong, 2011).

Fiberglas™ is the most common fiber-reinforced composite. For making its matrix, polyester is reacted with double bonds of carbon-carbon and styrene. A mix of polyester and styrene is poured over glass fibers (Zhao and Gou, 2009).

The double bonds present in the polyester and styrene react by free radical vinyl polymerization for forming a crosslinked resin. Inside, the glass fibers act as a reinforcement, and they are actually trapped there. In any particular direction, the fibers in Fiberglas™ are not lined up and appear like a tangled mass. The composite can be made stronger by lining the fibers in the same orientation. Some weird things are done by oriented fibers to the composite. When the composite is pulled in the fibers' direction, the composite is very strong. If they are pulled at the angle of 90° to the fiber direction, the composite is not strong (Prokhorov, 1970). Sometimes it is not bad if the composite is required to be strong in a single direction. The item that is being designed might be under stress in a single direction. On the other hand, sometimes, strength is required in various directions. In this case, the fibers are pointed in those required directions. This can be done by making use of fiber's woven fabric for reinforcing the composite. Good strength is provided by these woven fibers in many directions (Jassal and Ghosh, 2002; Gabara, 2006).

The fibers are held together by the polymeric matrix. Fibers loose bundle is not useful. Also, fibers can be brittle (regardless of how strong they are). If the matrix deforms under stress, it can absorb energy. Toughness is added to the composite through the matrix. Moreover, fibers have awful compressional strength, while having good tensile strength. When squashed, the fibers buckle. To the composite, compressional strength is provided by the matrix (Giancaspro et al., 2009).

All fibers are not the same. It may sound strange that though it is easy to break the glass, still it is used as reinforcement. But if the glass is spun in fibers (tiny ones), for some reason, the glass acts very differently. This type of glass fiber is flexible and strong. Even more strong fibers are present. This is advantageous as sometimes; glass just is not tough and strong enough. For some airplane parts like things, which can undergo massive stress, there is a need of breaking out the fancy fibers (García et al., 2010).

If cost is not a limiting factor, expensive, and stronger fibers, like Kevlar™, can be used. In contrast to Kevlar™, carbon fiber, Spectra™, is comparatively stronger. Carbon fiber is capable of withstanding greater force without breaking. Kevlar™ is tougher, which means it is capable of absorbing greater energy without breaking. Kevlar™ can stretch a little, which prevents it from breaking. But Spectra™ is tougher and stronger than both Kevlar™ and carbon fiber (Smallman and Bishop, 1999).

3.3.2. Biocomposites

Timber has been used by the residential construction field for many decades. Timber serves as an important source of building material in modern American homes. In 1999, 49.5 billion board feet of lumber were produced by the American timber industry and in 2002, 48.0 billion board feet were produced. The lumber production is peaking at the same time. In 2002, with more than 977,000 homes being sold, a record high homeownership rate of 69.2% was observed. In the United States, due to the excessive use of softwood lumber, the demand for alternate materials is increasing. Using sawdust fulfills this demand and also, the by-product is used efficiently (Figure 3.5) (Alizadeh-Osgouei et al., 2019; Miyagi and Teramoto, 2019).

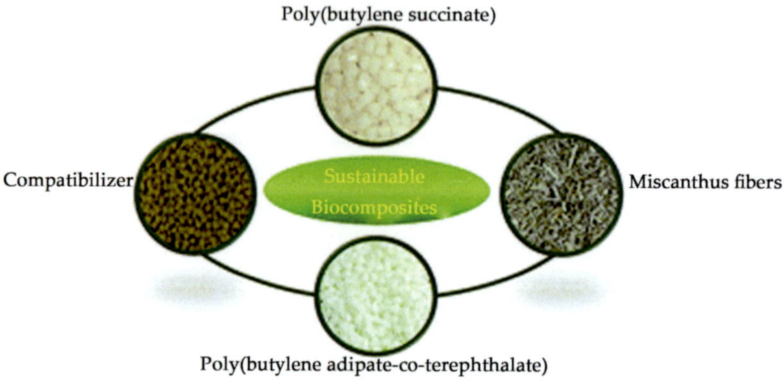

Figure 3.5. Different raw materials used to synthesize the biocomposites.

Source: https://pubs.rsc.org/en/content/articlelanding/2017/ra/c6ra27987b#!divAbstract.

A new category of materials, named wood-plastic composites (WPC), covers a wide range of composite materials, making use of fillers (that are composed of cellulose materials) and organic resin binder (matrix). The rapidly developing and new biocomposite materials are high technology products that have the advantage that scrap wood and sawdust products can be included in the wood filler. Any additional wood resources are not required for manufacturing biocomposites (Partap et al., 2007). Waste products may traditionally require a lot of money for proper disposal, but has turned out to be a beneficial resource, as it allows recycling that makes it both environmentally conscious and profitable. All over the world, the use of WPC and biocomposites has increased rapidly. These composites have their end-users in the furniture, motor vehicle, and construction industries. While using biocomposites, the major problem is the flammability of both components, i.e., filler, and binder. In case, flame retardants are added, it will be required that the retardant does not disturb the adhesion of the matrix and fiber. Developing a composite, which maintains its mechanical performance level and will not burn, is challenging. As an alternative to organic matrix compounds, we can use inorganic matrices for improving fire resistance. Wood composites (inorganic-based) contain a mineral mix that acts as the binder system. These systems (inorganic binder systems) include Portland cement and gypsum, both are highly resistant to insects and fire. These systems have a major disadvantage that a very low amount of fibers or sawdust can be incorporated by them. Potassium aluminosilicate is a new type of inorganic matrix. It is formed from naturally occurring materials and is environmentally friendly. The feasibility of the usage of this matrix, investigated by the Federal Aviation Administration, in commercial aircraft because of its ability to offer resistance to temperatures of up to 1000°C (while not generating the smoke), and its ability to enable carbon composites to withstand 800°C temperatures and to maintain its original flexural strength up to 63%. Potassium aluminosilicate matrices are suitable for many building materials such as masonry, particleboard, clay brick, steel, titanium, pine, concrete, balsa, and oak (Auborg et al., 1991). Two distinct processes are involved in FRP. In the first process, the fibrous material is formed and manufactured, and in the second, the fibrous materials are bonded (during the molding process) with the matrix (Rogovina, 2016).

3.4. FIBER MANUFACTURING PROCESS

Techniques that are included in fiber manufacturing processes are discussed in subsection (Šimáček and Advani, 2004; Park et al., 2011).

3.4.1. The Fiber Fabric Manufacturing

In both, two, and three-dimensional orientations, reinforcing fiber is manufactured. A laminated structure characterizes the two-dimensional fiber-reinforced polymer. In this orientation, the fibers are aligned, in y-direction and x-direction of the material, along the plane. Thus, along the z-direction or the through-thickness, no fibers are aligned. This lack of alignment may result in a disadvantage in processing and cost. Labor and cost increases as the conventional processing techniques for fabricating composites, such as resin transfer molding (RTM), autoclave, and hand lay-up, require highly skilled labor for cutting, stacking, and consolidating into a preformed component (Trochu et al., 2006).

3.4.2. The Fiber Preforms Manufacturing

Before the fibers are bonded to the matrix, the way they are manufactured is called fiber performs. They are manufactured in continuous mats, sheets, or as continuous filaments. For manufacturing, the fiber preforms, the major way is by the means of the textile processing techniques of braiding, weaving, stitching, and knitting (Chen and Chou, 1999).

Weaving can be performed in a traditional manner for producing two-dimensional fibers and for multilayer weaving, which is used to create three-dimensional fibers. However, multiple layers of warp yarns are required for creating fibers in the z-direction, which may lead to disadvantages in manufacturing. For producing high-value products (keeping the cost of producing preform in control) or relatively narrow width products, multilayer weaving is currently used.

For manufacturing fiber preforms, the second major way is braiding. Braiding is suitable in manufacturing tubular or narrow width flat fabric. Though for producing large volumes of wide fabrics, weaving is more capable still, weaving can be used. Braiding is done excessively on top of mandrels that vary in dimension along their length or cross-sectional shape. It is limited to objects. Unlike the standard processes of weaving, braiding can produce fabrics containing fibers, which have an angle of 45° to one another. By using Multilayer Interlock, two-step or four steps Braiding, braiding three-dimensional fibers are done. A flatbed is utilized by four-step or column and row braiding that contains columns and rows of yarn carriers for forming the shape of the preform. Outside of the array, additional carriers are added. The quantity and precise location of carriers depend upon the exact structure and shape of the required preform. Column and row motion has four separate sequences that act to produce the braided preform and interlock the yarns. Mechanically, the yarns are forced towards the structure among each step for consolidating the structure into alike processes. As compared to the four-step process, two-step braiding is different as it includes a lesser number of braiding yarns, and a huge number of yarns are fixed in the axial direction. Through the structure, the braiding carriers move between the axial carriers. This sequence of motions can form preforms of different shapes, such as hollow and circular. This two steps process is dissimilar to the four steps process in another way, as mechanical compaction is not required in it. In the process, yarn tension alone can tightly pull the braid. Multi-layer interlocking braiding is the last type of braiding. It contains numerous standard circular braiders that are joined together for forming a cylindrical braiding frame. Around the cylinder's circumference, numerous parallel braiding tracks are present in this frame. However, the transfer of yarn carriers is allowed by this mechanism, which forms a multilayer braided fabric having yarns interlocked to adjacent layers (Figure 3.6) (Campbell et al., 2008).

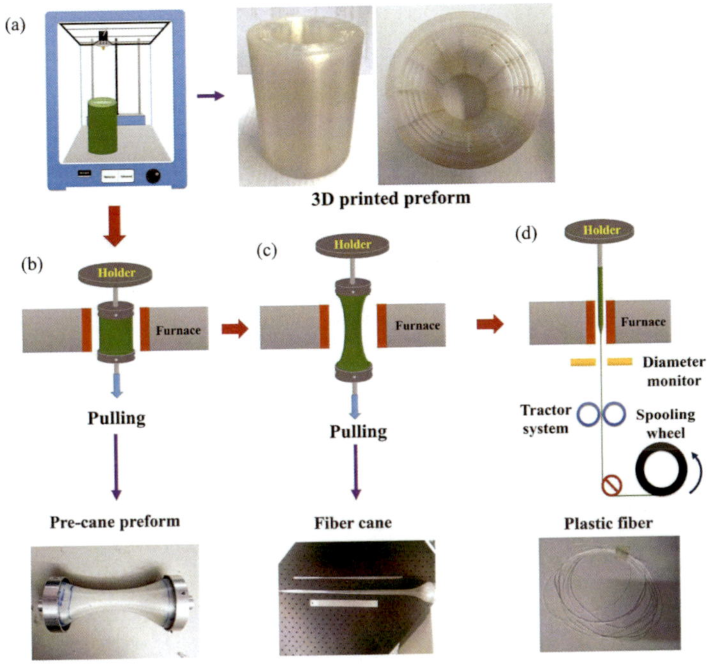

Figure 3.6. Fiber preforms: (a) a commercial 3D printer used to print a hollow-core fiber preform; (b) high-temperature pre-caning for reducing the size of the preform followed by; (c) cane drawing; and (d) fiber drawing.

Source: https://www.researchgate.net/figure/a-Hollow-core-fiber-preform-printed-using-a-commercial-3D-printer-b-High-temperature_fig1_325368176.

The multilayer interlock braid is different from both the two-step and four-step braids, as in multilayer interlock braid, the interlocking yarns are present mainly in the plane and hence, the in-plane properties of the preform are not reduced significantly. In contrast to this, a greater degree of interlinking is produced by the two-step and four-step processes (Pillai, 2004). This is because, in both these processes, the braiding yarns move via the thickness of the preform, though, this results in less contribution to the preforms in-plane performance. The multilayer interlock equipment has the disadvantage that the equipment is unable to possess the density of yarn carriers. This disadvantage is because of the yarn carriers' conventional sinusoidal movement (for the formation of preforms).

Conventional methods of Warp and Knitting [Weft] can be used for knitting fiber preforms. The resultant fabric can be regarded as a two-dimensional fabric. If machines have more than one needle bed, multilayer fabrics (which can traverse between the layers) can be produced. Developments in electronic controls for knit loop transfer and needle selection are being made. Also, in the sophisticated mechanisms, which allow to hold specific areas of the fabric and controls their movement, developments are being made. This allows the formation of fabric into the desired three-dimensional perform shape (having negligible material wastage).

Among the main textile manufacturing techniques, stitching is the simplest. Stitching techniques can be carried out with minimum possible investment in specialized machinery. This process involves the insertion of a needle that through a stack of fabric layers, carries stitch thread for forming a 3D structure. This stitching technique has the advantage that it allows stitching both prepreg and dry fabric, although the process becomes difficult because of the tackiness of prepare and generally, as compared to the dry fabric, it results in more damage within the material (prepreg).

3.5. COMPOSITE MOLDING

Wet molding and composite molding are two different categories of molding processes that use FRP plastics. Wet molding combines the matrix and fiber reinforcement or resists during the process. Composite molding makes use of Prepreg FRP, which means that before going through further molding processes, the plastics are fiber-reinforced. Prepreg FRP sheets are compressed or heated in different ways for creating geometric shapes. The different forms of wet molding and composite molding are discussed in subsections (Rusnakova et al., 2016; Schillfahrt et al., 2017).

3.5.1. Bladder Molding

Along with a bladder (balloon-like) prepreg material's individual sheets are placed and laid-up in a female-style mold. In a heated press, the mold in closed form is placed. Then, the bladder is pressurized. This pressure forces the material's layers against the mold walls. Once the layers are pressurized, the part is cured, after which, it is removed from the hot mold. The process of bladder molding has a short cure cycle that ranges from 15 to 60 minutes. This short cure cycle makes it ideal for forming complex hollow geometric shapes at a reasonable price (Figure 3.7) (Colton and Beeson, 1996; Anderson and Altan, 2012).

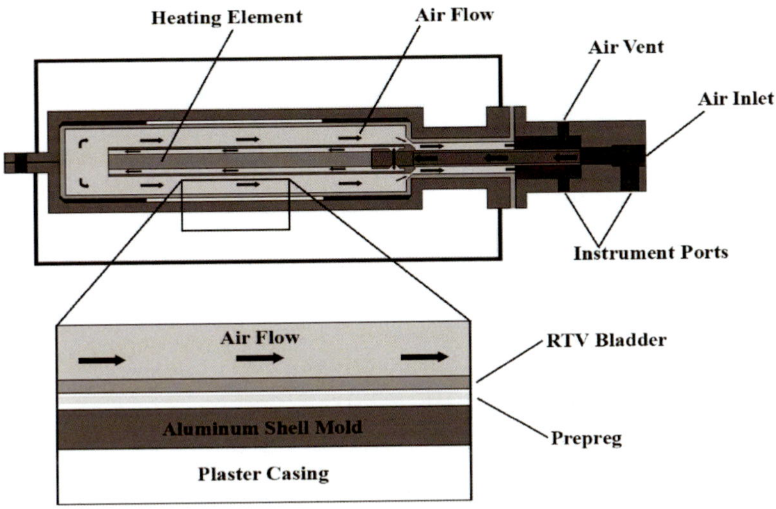

Figure 3.7. Bladder assisted composite manufacturing (BACM) fabrication assembly along with the process schematic.

Source: https://www.researchgate.net/figure/Bladder-Assisted-Composite-Manufacturing-BACM-fabrication-assembly-and-process_fig3_268513577.

3.5.2. Compression Molding

A "charge" or "preform," of prepreg fabric, BMC, or SMC, is put into a mold cavity. The material is cured and compacted inside by heat and pressure and the mold is closed. For geometric shapes, excellent detailing is offered by compression molding (Lee et al., 1984; Yi and Jain, 2005).

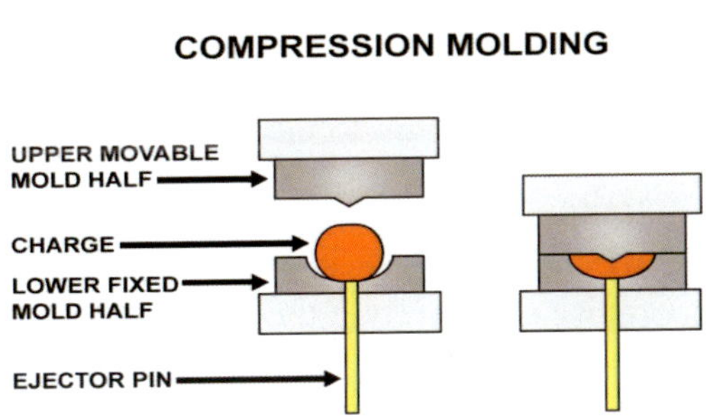

Figure 3.8. Compression molding illustrated schematically.

Source: https://www.martins-rubber.co.uk/blog/advantages-and-disadvantages-of-compression-molding/.

Compression molding provides detailing for a wide range of geometric shapes that include relief detailing, pattern detailing, creative forms, complex curves, and others having precision engineering within 20 minutes (curing time) (Figure 3.8).

3.5.3. Autoclave-Vacuum Bag

In an open mold, prepreg material's individual sheets are placed and laid up. The material is covered with a vacuum bag, bleeder/breather material, and release film. The mold is laid down into an autoclave that is a heated pressure vessel, and a vacuum is pulled on the part (Figure 3.9) (Barone and Caulk, 1986; Advani and Tucker III, 1990).

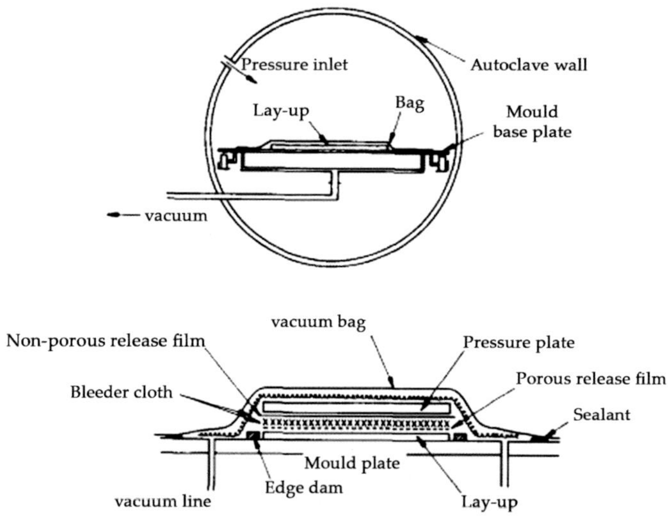

Figure 3.9. Vacuum bagging autoclave technique.

Source: https://www.sciencedirect.com/topics/engineering/vacuum-bag.

For extracting entrapped gasses from laminate, a continuous vacuum is used to cure the part. This process is commonly used in the aerospace industry, as it offers precise control over the molding process that is because of the slow and long cure cycle, which may take up to two hours. The required laminate geometric forms are created with the help of this precise control. The exact forms are necessarily required as they ensure safety and strength in the aerospace industry. Although, the labs are costly and the process is slow, but it is suitable for the aerospace industry (Hou and Jensen, 2004; Park et al., 2019).

3.5.4. Mandrel Wrapping

Prepreg material's sheets are wrapped around an aluminum or a steel mandrel. Polypropylene (PP) or nylon cello tape compacts the prepreg material. Parts can be batch cured in an oven (hanging in an oven). The mandrel and cello are removed after the cure, which leaves behind a hollow carbon tube. Robust and strong hollow carbon tubes are created through this process (Figure 3.10) (Cherin et al., 1982; Stringer, 1989).

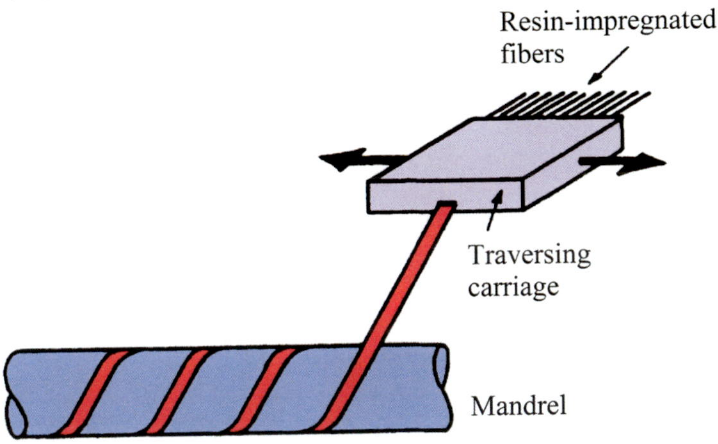

Figure 3.10. The composite manufacturing process of the mandrel wrapping.

Source: https://link.springer.com/chapter/10.1007/978-3-030-28983-6_5.

3.5.5. Wet Layup

Fiber reinforcing fabric is laid down in an open mold. Then, by pouring a wet resin over the fabric, it is saturated and molded. The mold is then left at room temperature so that the resin can cure (sometimes heat is used for ensuring a proper curing process). For this process, glass fibers are mostly used, the resultant products are known as fiberglass. Fiberglass is widely used in the manufacturing of common products such as surfboards, skis, kayaks, and canoes (Liu et al., 2011; Thompson et al., 2012).

3.5.6. Chopper Gun

Strand of fiberglass is continuously forced through a hand-held gun. This gun, after chopping, combines the strands with a catalyzed resin, i.e., polyester. The human operator shots the impregnated chopped glass onto the mold surface in the design and thickness that he thinks is appropriate. For large production runs, this is a useful process as it is economical. A major disadvantage of this process is that the produced geometric shapes have less strength and comparatively poor dimensional tolerance (Abanilla et al., 2005; Karbhari and Abanilla, 2007).

3.5.7. Filament Winding

Fiber bundles are pulled by the machines through a wet resin bath and then over a rotating steel mandrel; they are wounded in specific orientations. At elevated temperatures or room temperature, the parts are cured and the mandrel is extracted, which leaves behind a final geometric shape (Figure 3.11) (Jiao et al., 2012; Banibayat and Patnaik, 2014).

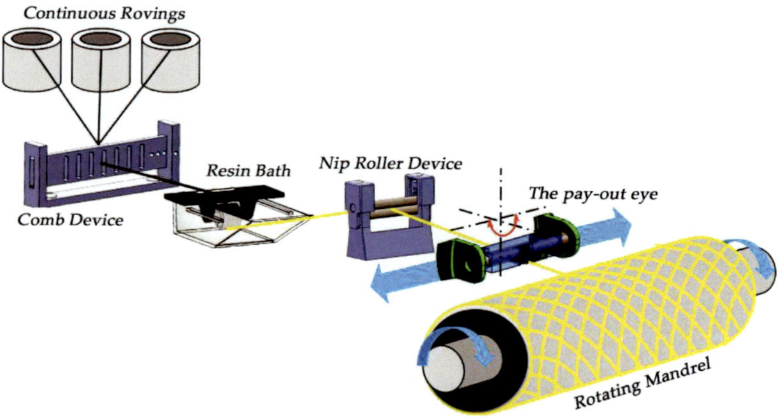

Figure 3.11. Filament winding process.

Source: https://www.researchgate.net/figure/Schematic-of-filament-winding-process_fig1_332329420.

3.5.8. Pultrusion

Slit fabrics and fire bundles are pulled through a wet resin bath and then are formed into the rough shape of the part. From a closed heated die curing, saturated material is extruded. The resultant products of this process are structural shapes such as flat sheets, beam, channel, and angle. These products are used in manufacturing all kinds of fiberglass structures such as handrail systems tanks, pump supports, ladders, pipes, and platforms (Figure 3.12) (Abanilla et al., 2006; Centea and Nutt, 2016).

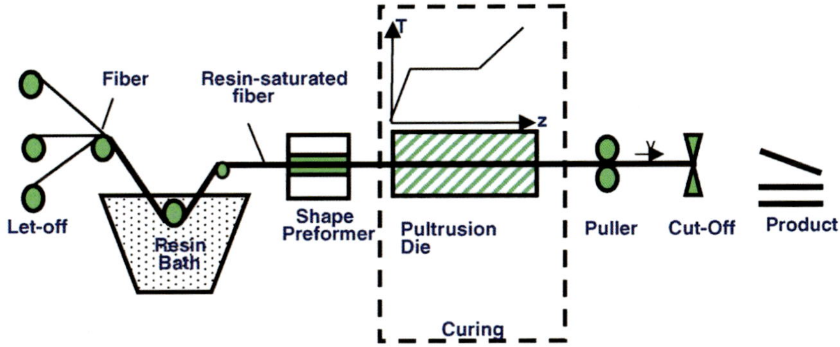

Figure 3.12. The pultrusion process illustrated schematically.

Source: https://www.researchgate.net/figure/Schematic-of-the-pultrusion-process-for-fabrication-of-thermosetting-matrix-composites_fig1_223914700.

3.6. RESIN INFUSION

Fabrics are laid down into a mold first and then, the wet resin is injected. Resin is typically forced and pressurized into a cavity that is present under vacuum in the RTM process. Into the cavity, the resin is pulled under vacuum in the vacuum-assisted resin transfer molding (VARTM) process. Detailed shaping and precise tolerances are allowed by this molding process. Though, sometimes this process cannot fully saturate the fabric that results in weak spots in the end product (Figure 3.13) (Chen et al., 2004a, b; Morshed et al., 2019).

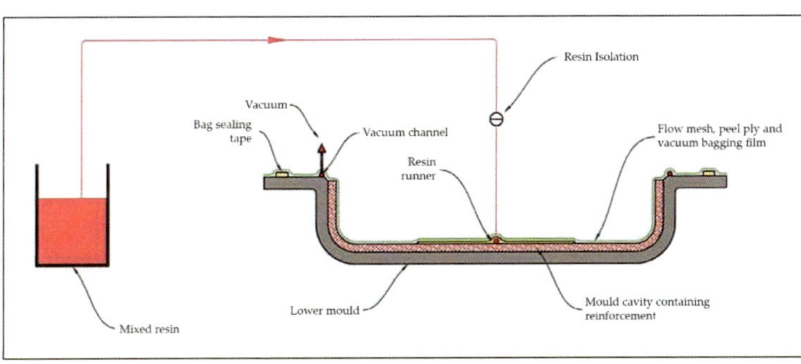

Figure 3.13. Vacuum resin infusion process.

Source:https://www.researchgate.net/figure/Schematic-diagram-of-vacuum-resin-infusion-process_fig3_309593729.

3.6.1. Advantages and Limitations

The glass fibers alignment of thermoplastics is allowed by FRP. This alignment complements specific design programs. Specifying the reinforcing fiber's orientation, the resistance to deformation and strength of the polymer can be increased. Glass reinforced polymers are extremely resistive to deforming forces and are strongest, especially when the polymers fibers and the force being exerted are parallel. In case, when the polymer fibers and the force being exerted are perpendiculars, glass-reinforced polymers are weakest. Hence, depending on the usage, this ability can serve as a limitation, as well as an advantage (Crandall and Hartle, 1985). We can use the weak spots of perpendicular fibers for connections and natural hinges, but can also result in material failure in a case if production processes fail in parallelly orienting the fibers against the expected forces. When forces act perpendicular to the fiber's orientation, the elasticity and strength of the polymer are relatively less (in contrast to the matrix alone). In glass-reinforced polymers constituted cast resin components, such as EP and UP, fibers can be oriented in three-dimensional and two-dimensional weaves. We can conclude that when forces are acting perpendicular to one orientation, possibly, they might be parallel to another orientation; eliminating the weak spots in the polymer (Devries et al., 2001).

3.6.2. Failure Modes

In FRP materials, structural failure can occur when (Peters et al., 1991; Mertiny and Ellyin, 2002):

- The matrix is stretched more by the tensile forces as compared to the fibers. This results in the material shearing at the interface between fibers and matrix.
- Near the fiber's end, the tolerances of the matrix are exceeded by the tensile forces, which results in the separation of the fibers from the matrix.
- The tolerances of the fibers can also be exceeded by the tensile forces, which causes the fibers to fracture that ultimately results in material failure (Erhard, 2013).

3.6.3. Material Requirements

Certain requirements must be met by the matrix in order to complement the FRP process and for ensuring its successful reinforcement. The matrix should saturate properly. The matrix should form bonds with the fibers in a promising curing period. It should also form chemical bonds with the fiber reinforcement to ensure maximum adhesion. The fibers should be completely enveloped by the matrix, as it is important for protecting them from notches and cuts that may reduce their strength (Van de Velde and Kiekens, 2001; Colombo and Vergani, 2018). To avoid complete failure, the fibers should be placed separately from each other and in case of failure takes place, the matrix should also debond from the fiber. The matrix should be preferably plastic, as it remains physically and chemically stable after (and during) molding and reinforcement processes. The modulus of elasticity and tensile strength of the matrix must be increased to make them suitable for fiber (reinforcement material) additives. Similarly, the following conditions should also be met; critical fiber content must be exceeded by the fibers; the rigidity and strength of fibers itself should exceed the rigidity and strength of the matrix alone, and optimum bonding must be there between matrix and fibers (Han et al., 1986; Carlone et al., 2006).

3.7. GLASS FIBER MATERIAL

Textile glass fibers are used by FRPs; in contrast to other glass fibers that are used for insulating applications, textile fibers are different. Textile glass fibers are initially in varying combinations of Al_2O_3, CaO, SiO_2, MgO, or B_2O_3. Then, a direct melt process is used to heat these mixtures at 1300°C. Afterward, dies are used for extruding glass fiber filaments that in diameter ranges between 9 and 17 μm. After this, filaments are spun onto bobbins (after being wounded into larger threads) for further processing and transportation (Miller et al., 1998). The most popular means for reinforcing plastic is a glass fiber, and thus, it undergoes different production processes. Among these processes, some are applicable to carbon fibers and aramid. In a process named 'roving,' filaments are spun into threads that have a larger diameter. Larger diameter threads are used in spray applications, and for woven reinforcing mats and glass fabrics. Fiber fabrics are reinforcing fabric (web-form) material, having both weft and warped directions. Fiber mats are non-woven web-form mats of glass fibers. They are manufactured with chopped fibers in cut dimensions, or by using continuous fibers in continuous mats. In processes where the glass threads are cut in a length between 3 and 26 mm, chopped fiberglass is used. These threads are used in plastics (mostly for molding processes). Short strands of glass fiber are 0.2–0.3 mm. They are used for reinforcing thermoplastics usually for injection molding (Williams et al., 1996; Liu et al., 2000).

3.8. ARAMID FIBER MATERIAL PROCESS

Aramid fibers are usually known as Technora, Nomex, and Kevlar. Generally, aramids are prepared by reacting an aramid, which is a carboxylic acid halide group, and an amine group; commonly, this reaction takes place when an aromatic polyamide (PA) is spun into a crystallized fiber from sulfuric acid's liquid concentration. In order for weaving into woven fabrics or large ropes, fibers are spun into larger threads. With varying grades, aramid fibers are manufactured. They base on varying qualities for rigidity and strength so that it is possible to tailor the material according to desired specific design concerns, i.e., cutting of the hard material while manufacturing (Goren and Atas, 2008; Govignon et al., 2010).

3.9. FRP APPLICATIONS

FRPs are excellent for design programs that require finite tolerances, precision engineering, weight savings, and parts simplification in both operation and production. In contrast to steel artifact or cast aluminum, a molded polymer artifact is easier, cheaper, and faster to manufacture. A molded polymer artifact maintains better material strengths and tolerances. FRP was also used by the Mitsubishi Lancer Evolution IV for its spoiler material (Rosato and Rosato, 2004; Schell et al., 2007).

3.9.1. Carbon Fiber Reinforced Polymers

Carbon fiber reinforced polymers are used to manufacture the rudder of a commercial airplane. As compared to a traditional sheet aluminum rudder, it has the following advantages (Joubaud et al., 2005; Niggemann et al., 2008):

- Around 25% weight reduction;
- Around 95% components reduction, as it combines different parts, forming them into simpler molded parts;
- Overall reduction in operational and production costs, the economy of parts leads to lower production costs and less weight saves fuel, resulting in lower operational costs.

3.9.2. FRP's Structural Applications

FRP can be applied for strengthening the slabs, columns, and beams in buildings. The strength of these mentioned structural members can be increased even if they have been severely damaged. Two different techniques are usually adopted to strengthen the beams. In the first technique, FRP plates are placed at the bottom of a beam (usually the tension face). This increases the deflection capacity of the beam, the strength of the beam, and stiffness. On the other hand, we can place FRP strips around the sides (and at the bottom) in a 'U' shape. This provides higher shear resistance. FRP can be used to wrap the columns in the building. This process is termed as called wrapping of columns, it helps in achieving higher strength. Pasting FRP strips at the bottom of the slabs (tension face) strengthens them. Pasting strips at the bottom results in better performance, as the tensile strength of FRP supplements the tensile resistance of slabs. In the case of slabs and beams, the effectiveness of FRP strengthening is dependent on the resin chosen for bonding, the better will be the performance of the resin, the better will be the effectiveness (Klompen, 2005; Kumar and Sundaram, 2016).

3.9.3. Glass Fiber Reinforced Polymers

Glass fiber reinforced PA 66 is used to make engine intake manifolds. Over cast aluminum manifolds, engine intake manifolds have the following advantages (Gupta and Sundaram, 2009):

- Around 60% weight reduction;
- Improved aerodynamics and surface quality;
- Components reduction by combining different parts, forming them into simpler molded shapes. Glass fiber reinforced PA 66 is used for making clutch pedals and automotive gas (DWP 12–13).
- Following are the advantages it has over stamped aluminum:
- We can mold the pedals as single units combining mechanical linkages and pedals simplifying the production and design operation;
- Fibers can be oriented for reinforcing them against specific stresses. This increases safety and durability.

3.9.4. Design Considerations

Designs, which require modulus of elasticity or measure of the strength of nonreinforced plastics (even other material choices) are economically or mechanically unsuitable. It can be concluded that for using FRP, the primary design consideration is to make sure that the material being used is economical and it should be used in a way that complete advantage of its structural enhancements is taken. However, the case is not always the same, the orientation of fibers may create a material weakness. Thus, the orientation and use of fiber affect the elasticity, rigidity, and strength of a final form and also its operation (Peck, 1969; Liu and Xia, 2010). During production, 3-dimensionally, 2-dimensionally, or unidirectionally (UD) orienting the direction of fibers affects the elasticity, flexibility, and strength of a final form. A greater resistance, by the fibers that are oriented in the same direction as of forces, to distortion from forces is observed (vice versa). Hence, the product area, which should withstand forces, has to be reinforced with fibers in a similar direction, whereas the areas for which flexibility is important (for example, natural hinges) can use fibers in a direction that is perpendicular to forces. This can be avoided if more dimensions are used, this creates objects that do not have any specific weak points. The properties of elasticity, flexibility, and strength can also be diminished or magnified through the design and geometric shape of the end product. These include design considerations, for instance creating multifunctional geometric shapes and ensuring proper wall thickness, that can be molded in the form of single pieces, manufacturing shapes that have more structural integrity and more material by reducing hardware, connections, and joints (Rosato and Rosato, 2004; Lefever et al., 2007).

3.9.5. Recycling and Disposal Concerns

FR plastics are a subset of plastic, so they are liable to numerous concerns and issues in plastic waste recycling and disposal. In recycling processes, a particular challenge is posed by plastics as they are derived from monomers and polymers, it is impossible to return them to their virgin states or to separate them, thus all plastics cannot be recycled for re-use. According to some estimates, the percentage of plastics that can be material recycled is between 20% and 30%. These environmental and disposal concerns are shared by the FR plastics and their matrices. Moreover, it is difficult to remove fiber (themselves) from the matrix. Also, it is hard to preserve them for re-use means. Inherently, separating FRPs into base materials (matrix and fiber) is difficult. Similarly, separating FRP matrix into monomers, polymers, and plastic is difficult. These all concerns regarding the environment are major ones, but on the other hand, in contrast to other materials, plastics also offer economic savings and energy saving. Especially the environmentally friendly matrices, for instance, UV-degradable plastics and UV-degradable plastics, will gain environmental sensitivity (Pearson, 2002; Erhard, 2013).

3.10. MECHANICAL PROPERTIES MEASUREMENTS

3.10.1. Strength

As the word strength is more related to a mechanical property, but here we do not have an idea of what the word strong may mean when we are discussing polymers. We need to understand that more than one kind of strength is present. Firstly, we will be considering tensile strength. If a polymer is strong when it is being pulled, it has tensile strength. For a material, this strength is important if it is under tension or is being stretched. Good tensile strength is required by the fibers (Zulfiqar et al., 2019).

Then comes compressional strength. If a polymer sample is strong when it is being compressed, it has compressional strength, for example, concrete. Things that need to support weight, must have good compressional strength (Klompen, 2005). Flexural strength is another kind of strength. If a polymer sample is strong when it is being bent, it has flexural strength. Other kinds of strength are also there. If a polymer sample is strong when it is being twisted, it has torsional strength. Then comes impact strength. If a polymer sample is strong when it is being hit suddenly and sharply (like the blow from a hammer), it has impact strength (Bull, 1991).

For measuring the tensile strength of a sample, we try to stretch the sample, usually with machines. These stretching machines have clamps on the corners of the sample. Turning on the machine stretches the sample. The amount of force (F) being exerted is measured when the sample is stretching. The amount of force is then divided by the sample's cross-sectional area (A). This gives us the total stress that is being experienced by the sample. Until the sample breaks, the amount of stress and force is continuously increased through the machine. The tensile strength of the sample is the stress that is required to break it. Similar tests for flexural or compressional strength can be imagined. Tensile strength and tensile stress, are measured usually in N/cm^2. Strength and stress can be measured in gigapascals (GPa) or megapascals (MPa). Different units can be converted easily, as 1 GPa = 1,000 MPa, 1 GPa = 100,000 N/cm^2, and 1 MPa = 100 N/cm^2. Other times, strength, and stress are measured in pounds per square inch (psi). The conversion factor for converting psi to N/cm^2 is 1 N/cm^2 = 1.45 psi (Arabian-Hoseynabadi et al., 2010).

3.10.2. Elongation

Knowing the strongness of a polymer is not enough, there is more to understand about its mechanical properties. Strength does not tell us anything except the stress required to break it. It does not give us an idea of what actually happened to the sample before it got broken. That is the point where the elongation behavior needs to be studied. We can say that elongation is a kind of deformation. By deformation, we simply mean the change in shape due to the applied stress. When discussing tensile stress, stretching deforms the sample and it becomes longer. Of course, it is termed elongation. Mostly, percent elongation is discussed. Percent elongation is the length of the sample after being stretched (L), divided by the sample length (L_0) (original length), and then multiplied by 100 (Whybark and Williams, 1976).

Numerous things related to elongation are measured. Most importantly, the type of material being studied depends a lot. Elastic elongation and ultimate elongation are two main things we measure. Ultimate elongation is the amount to which we can stretch the sample and hence is important for all kinds of materials. Elastic elongation is the maximum percent elongation that can be reached before permanently deforming the sample. We can say that this is the maximum amount by which we can stretch the material, and after being stretched when stress is released, it snaps back to its original length. In case the material is an elastomer, elastic elongation is important. Elastomers have to stretch a long-distance and after the release of stress, they bounce back. Mostly, they can stretch between 500% and 1000% elongation (Lindhagen and Berglund, 1998; Guo, 2007).

3.10.3. Modulus

In the elastomers, it is important to show high elastic elongation, while other types of materials, such as plastic, do not deform or stretch easily. In order to measure that to what extent material resists deformation, the modulus is measured. For measuring tensile modulus, the same thing is done as with the case of measuring ultimate elongation and strength. This time, the stress being exerted (by us) on the material is measured. Initially, we slowly increase the amount of stress. Afterward, at each stress level, the elongation the sample undergoes is measured. Until the sample breaks, we keep measuring the elongation (Crowson et al., 1980). This plot is termed as a stress-strain curve (any kind of deformation is referred to as strain (elongation is also a kind of deformation), whereas by elongation, we specifically mean tensile strain). The tensile strength is the height of the curve (at the point where the sample breaks). The slope of the plot is the tensile modulus. A steep slope suggests a high tensile modulus that means deformation is resisted. A gentle slope suggests a low tensile modulus that means a material can be easily deformed. Sometimes the stress-strain curve is not straight and nice. As stress is increased, the slope does not remain constant. With stress, the modulus is changing. In such cases, the changing modulus changes the initial slope (Azmir et al., 2009; Zhu et al., 2010).

In general, elastomers have the lowest tensile moduli, fibers have the highest, and plastics have somewhere in between elastomers and fibers.

We can measure the modulus by calculating stress and then, dividing it by elongation. Due to the fact that elongation has no units and is dimensionless, the modulus is measured in the unit of strength that is N/cm^2.

During homogeneous deformation, the true stress-strain response of the material is called intrinsic deformation. As strain localization phenomena take place (like shear banding, necking, cracking, and crazing), the intrinsic materials' response measurement needs a special experimental set-up, i.e., a uniaxial compression test or a video-controlled tensile test. The intrinsic response details may be different for each material, still, a general representation can be recognized in Figure 3.14 (Ploeg and Claessen, 2015).

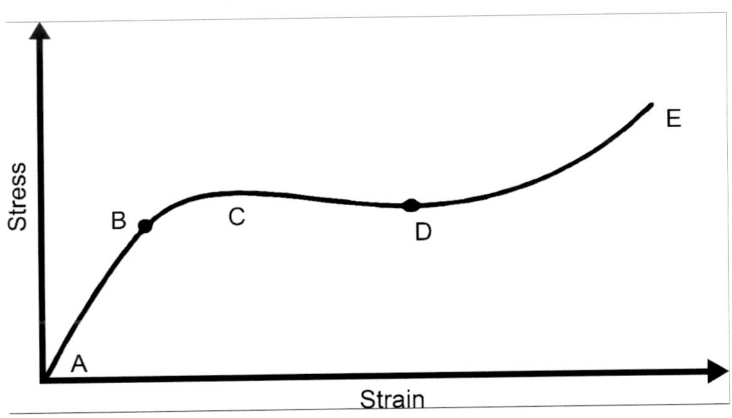

Figure 3.14. The polymer material's intrinsic deformation behavior schematically represented.

Source: https://en.wikipedia.org/wiki/File:Stress-Strain_Type_V.png.

3.10.4. Toughness

Another valuable piece of information is given by this plot of stress versus strain. If the area under the curve is measured, we get the value of toughness (Figure 3.15). In the graph given in Figure 3.16, it is colored red (Wu et al., 2007).

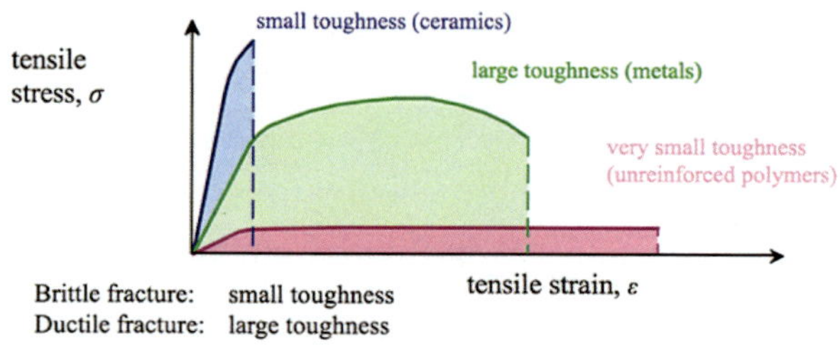

Figure 3.15. Stress-strain curve for a material that has varying toughness and tensile strength.

Source: https://www.e-education.psu.edu/matse81/node/2105.

Toughness is said to be a measure of the energy that a thing can absorb before breaking. If the base of the triangle is regarded as strain, and the height of the triangle as strength, its area would be proportional to strength strain. We measure the strain in n units of distance, as the strength and force required to break the sample are proportional. In this regard, strength strain and force times distance are proportional (according to physics, force times distance is the same as energy).

In the lights of physics, toughness is the energy required to break a sample and strength is the force required to break a sample. These definitions do not give a proper idea of the practical differences. The main point to remember is that even if a material is strong, it is not necessary for the material to be tough as well (Figure 3.16) (Odian, 1991; Jang, 1994).

For a sample that is not tough but is strong, the gray plot shows the stress-strain curve for it. As it can be observed that a lot of force is required for breaking the sample. Similarly, a sample cannot be stretched much before breaking. A material that cannot deform much before breaking and is strong is named brittle (Roy et al., 2019).

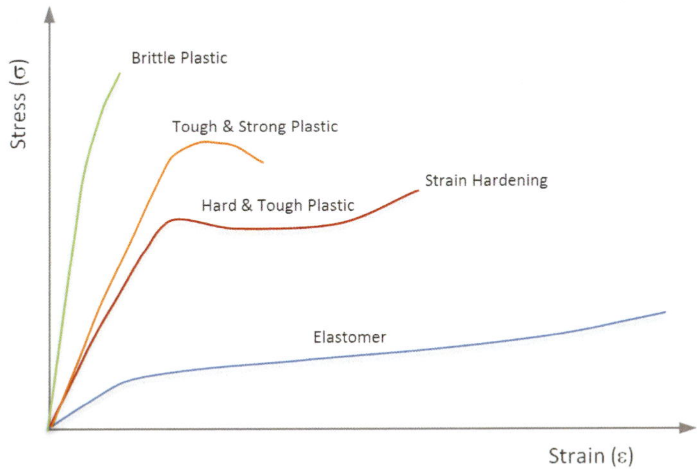

Figure 3.16. A stress-strain plot of tough and strong polymers.

Source: https://polymerdatabase.com/polymer%20physics/Stress-Strain%20 Behavior.html.

For a sample that is tough, as well as strong, the gray plot shows the stress-strain curve for it. This material is less strong than the one discussed above, but still, the area underneath its curve is comparatively larger. So, a lot more energy can be absorbed (Kim, 2019). The gray-black sample, in contrast to the gray sample, elongates a lot more. The deformation of a sample allows it to dissipate energy. Energy can never be dissipated if the sample does not deform, and this will result in the breaking of the sample (Hashim et al., 2011).

In daily life, we want materials to be strong and tough. Ideally, a material that does not break or bend would be preferred. In contrast to the red sample, the gray-black sample possess a higher modulus. In many applications, it is better for materials to resist deformation and have high moduli, as in daily life, it is much beneficial if a material bends rather than breaking. Similarly, deforming, stretching, or bending of the material is much better than its breaking. So, when designing new composites, or new polymers, strength is sometimes sacrificed to make the material tougher (Ruoff et al., 2003; Fery and Weinkamer, 2007).

3.11. REAL POLYMER'S MECHANICAL PROPERTIES

The rigid plastics, i.e., PC, polymethyl methacrylate, or polystyrene (PS) can bear greater stress, but they fail to withstand elongation and break. Under the stress-strain curve, there is very less area. We can conclude that such materials are strong, though they are not very tough. Moreover, the plot has a steep slope, which means that a lot of force is required for deforming a rigid plastic. Hence, rigid plastics have high moduli. It can be concluded that at resist deformation, gid plastics are strong, but they are brittle, as they are not very tough.

Flexible plastics like PP and PE are dissimilar to rigid plastics, as deformation is not resisted by them, but they try not to break. They do not break because of their ability to deform. Flexible plastics have a high initial modulus due to which they resist deformation for a while, but in case if enough stress is applied, flexible plastic deforms. At first, it will be hard to stretch flexible plastic like a plastic bag, but once stretched far, it will stretch easily. The main point is that flexible plastics might not appear to be strong and rigid, but they are much tougher.

Possibly, plastic's stress-strain behavior can be altered with additives named plasticizers. A small molecule that increases the flexibility of plastics is called a plasticizer. For instance, if a plasticizer is not used, PVC (polyvinyl chloride) is a rigid plastic. Unplasticized PVC is rigid, and that is why it is used as water pipes. PVC can be turned flexible, by using plasticizers, to make things like hoses.

Fibers like nylon, carbon fiber, and Kevlar™ tend to have aqua-colored plots like (the one shown in the above graph) stress-strain curves. They are stronger rather than being tougher (like the rigid plastics), and under tensile stress, they do not undergo deformation. As compared to plastics, fibers are much stronger. Moreover, some polymeric fibers, like ultra-high molecular weight PE, carbon fiber, and Kevlar™ have greater tensile strength than steel.

Elastomers like polyisobutylene, polybutadiene, and polyisoprene have different mechanical behavior. The above graph, looking at the pink plot, shows how their behavior is different from the other types of materials. They have very low moduli. This is depicted by the gentle slope in the graph. We are aware that it is easy to bend or stretch a piece of rubber. Low moduli are what make them very good elastomers.

REFERENCES

1. Aamir, M., Tolouei-Rad, M., Giasin, K., & Nosrati, A., (2019). Recent advances in drilling of carbon-fiber-reinforced polymers for aerospace applications: A review. *The International Journal of Advanced Manufacturing Technology, 105*(5), 2289–2308.
2. Abanilla, M. A., Karbhari, V. M., & Li, Y., (2006). Interlaminar and intralaminar durability characterization of wet layup carbon/epoxy used in external strengthening. *Composites Part B: Engineering, 37*(7, 8), 650–661.
3. Abanilla, M. A., Li, Y., & Karbhari, V. M., (2005). Durability characterization of wet layup graphite/epoxy composites used in external strengthening. *Composites Part B: Engineering, 37*(2, 3), 200–212.
4. Advani, S. G., & Tucker, III. C. L., (1990). A numerical simulation of short fiber orientation in compression molding. *Polymer Composites, 11*(3), 164–173.
5. Alizadeh-Osgouei, M., Li, Y., & Wen, C., (2019). A comprehensive review of biodegradable synthetic polymer-ceramic composites and their manufacture for biomedical applications. *Bioactive Materials, 4*, 22–36.
6. Al-Mutuki, A. P. S. M., Khalaf, L. B. A. H., & Abbas, L. A. A., (2016). Comparing study between the effect of wet and dry carbon fibers on hardness and impact of polystyrene carbon fiber sheets composite. *MJPS, 3*(2), 1–20.
7. Anderson, J. P., & Altan, M. C., (2012). Properties of composite cylinders fabricated by bladder assisted composite manufacturing. *Journal of Engineering Materials and Technology, 134*(4), 1–22.
8. Arabian-Hoseynabadi, H., Oraee, H., & Tavner, P. J., (2010). Failure modes and effects analysis (FMEA) for wind turbines. *International Journal of Electrical Power & Energy Systems, 32*(7), 817–824.
9. Aubourg, P. F., Crall, C., Hadley, J., Kaverman, R. D., & Miller, D. M., (1991). Glass fibers, ceramics, and glasses. *Engineered Materials Handbook, 4*, 1027–31.
10. Azmir, M. A., Ahsan, A. K., & Rahmah, A., (2009). Effect of abrasive water jet machining parameters on aramid fiber reinforced plastics composite. *International Journal of Material Forming, 2*(1), 37–44.

11. Babuska, I., & Oden, J. T., (2004). Verification and validation in computational engineering and science: Basic concepts. *Computer Methods in Applied Mechanics and Engineering, 193*(36), 4057–4066.
12. Banerjee, D., Nguyen, T., & Chuang, T. J., (2016). Mechanical properties of single-walled carbon nanotube reinforced polymer composites with varied interphase's modulus and thickness: A finite element analysis study. *Computational Materials Science, 114*, 209–218.
13. Banibayat, P., & Patnaik, A., (2014). Variability of mechanical properties of basalt fiber reinforced polymer bars manufactured by wet-layup method. *Materials & Design (1980–2015), 56*, 898–906.
14. Barone, M. R., & Caulk, D. A., (1986). *A Model for the Flow of a Chopped Fiber Reinforced Polymer Compound in Compression Molding, 1*, 1–20.
15. Bauhofer, W., & Kovacs, J. Z., (2009). A review and analysis of electrical percolation in carbon nanotube-polymer composites. *Composites Science and Technology, 69*(10), 1486–1498.
16. Borsato, M., (2017). An energy efficiency focused semantic information model for manufactured assemblies. *Journal of Cleaner Production, 140*, 1626–1643.
17. Boyer, R. F., (1968). Dependence of mechanical properties on molecular motion in polymers. *Polymer Engineering & Science, 8*(3), 161–185.
18. Boztepe, M. B., Bayramoglu, M., Uzay, C., & Geren, N., (2017). Investigation of the adhesion bonding capability between different metals and FRP composite layers. In: *Istanbul International Conference on Progress in Applied Science, ICPAS* (Vol. 1, pp. 1–6).
19. Bull, S. J., (1991). Failure modes in scratch adhesion testing. *Surface and Coatings Technology, 50*(1), 25–32.
20. Campbell, M., Denault, J., & Bureau, M. N., (2008). CF/PA12 composite femoral stems: Manufacturing and properties. *Composites Part A: Applied Science and Manufacturing, 39*(5), 796–804.
21. Carlone, P., Palazzo, G. S., & Pasquino, R., (2006). Pultrusion manufacturing process development by computational modeling and methods. *Mathematical and Computer Modeling, 44*(7, 8), 701–709.
22. Centea, T., & Nutt, S. R., (2016). Manufacturing cost relationships for vacuum bag-only prepreg processing. *Journal of Composite Materials, 50*(17), 2305–2321.

23. Chand, S., (2000). Review carbon fibers for composites. *Journal of Materials Science, 35*(6), 1303–1313.
24. Chen, B., & Chou, T. W., (1999). Compaction of woven-fabric preforms in liquid composite molding processes: Single-layer deformation. *Composites Science and Technology, 59*(10), 1519–1526.
25. Chen, H., Xi, N., Sheng, W., Chen, Y., & Dahl, J., (2004). Chopper gun trajectory optimization for spray forming in automotive manufacturing. In: *AIP Conference Proceedings* (Vol. 712, No. 1, pp. 1429–1434). American Institute of Physics.
26. Chen, H., Xi, N., Sheng, W., Chen, Y., & Dahl, J., (2004). Optimal spray gun trajectory planning with variational distribution for forming process. In: *IEEE International Conference on Robotics and Automation, 2004, Proceedings, ICRA'04* (Vol. 1, pp. 51–56). IEEE.
27. Cherin, A. H., Head, E. D., Lovelace, C. R., & Gardner, W. B., (1982). Selection of mandrel wrap mode filters for optical fiber loss measurements. *Fiber & Integrated Optics, 4*(1), 49–66.
28. Chow, T. S., (1980). The effect of particle shape on the mechanical properties of filled polymers. *Journal of Materials Science, 15*(8), 1873–1888.
29. Coleman, J. N., Khan, U., Blau, W. J., & Gun'ko, Y. K., (2006). Small but strong: A review of the mechanical properties of carbon nanotube-polymer composites. *Carbon, 44*(9), 1624–1652.
30. Colombo, C., & Vergani, L., (2018). Optimization of filament winding parameters for the design of a composite pipe. *Composites Part B: Engineering, 148*, 207–216.
31. Colton, J. S., & Beeson, M. S., (1996). Fiber coiling during bladder molding of thermoplastic composites. *Polymer Composites, 17*(4), 627–636.
32. Crandall, M. S., & Hartle, R. W., (1985). An analysis of exposure to styrene in the reinforced plastic boat-making industry. *American Journal of Industrial Medicine, 8*(3), 183–192.
33. Crowson, R. J., & Folkes, M. J., (1980). Rheology of short glass fiber-reinforced thermoplastics and its application to injection molding; II. The effect of material parameters. *Polymer Engineering & Science, 20*(14), 934–940.

34. Devries, J. E., Chavka, N. G., & Dahl, J. S., (2001). Recent advances in glass fiber preforming: Implementation of the ford programmable preform process (F3P). *Manufacturing of Advanced Composites, 21*, 1–22.
35. Erhard, G., (2013). *Designing with Plastics* (Vol. 1, pp. 1–20). Carl Hanser Verlag GmbH Co KG.
36. Erki, M. A., & Rizkalla, S. H., (1993). *FRP Reinforcement for Concrete Structures* (Vol. 15, pp. 40–48). Concrete International-Detroit.
37. Ferry, J. D., & Strella, S., (1958). Dielectric dispersion in methacrylate polymers and its correlation with mechanical properties. *Journal of Colloid Science, 13*(5), 459–471.
38. Fery, A., & Weinkamer, R., (2007). Mechanical properties of micro- and nanocapsules: Single-capsule measurements. *Polymer, 48*(25), 7221–7235.
39. Fu, S. Y., & Lauke, B., (1996). Effects of fiber length and fiber orientation distributions on the tensile strength of short-fiber-reinforced polymers. *Composites Science and Technology, 56*(10), 1179–1190.
40. Gabara, V., (2006). In: Lewin, M., (ed.), *Handbook of Fiber Chemistry* (pp. 1–22). International Fiber Science and Technology Series, Ch. 13.
41. García, J. M., García, F. C., Serna, F., & José, L. D. L. P., (2010). High-performance aromatic polyamides. *Progress in Polymer Science, 35*(5), 623–686.
42. Giancaspro, J., Papakonstantinou, C., & Balaguru, P., (2009). Mechanical behavior of fire-resistant biocomposite. *Composites Part B: Engineering, 40*(3), 206–211.
43. González, C., & LLorca, J., (2007). Mechanical behavior of unidirectional fiber-reinforced polymers under transverse compression: Microscopic mechanisms and modeling. *Composites Science and Technology, 67*(13), 2795–2806.
44. Goren, A., & Atas, C., (2008). Manufacturing of polymer matrix composites using vacuum assisted resin infusion molding. *Archives of Materials Science and Engineering, 34*(2), 117–120.
45. Govignon, Q., Bickerton, S., & Kelly, P. A., (2010). Simulation of the reinforcement compaction and resin flow during the complete resin infusion process. *Composites Part A: Applied Science and Manufacturing, 41*(1), 45–57.

46. Gullapalli, S., & Wong, M. S., (2011). Nanotechnology: A guide to nano-objects. *Chemical Engineering Progress, 107*(5), 28–32.
47. Guo, L. J., (2007). Nanoimprint lithography: Methods and material requirements. *Advanced Materials, 19*(4), 495–513.
48. Gupta, N., & Sundaram, R., (2009). Fiber optic sensors for monitoring flow in vacuum enhanced resin infusion technology (VERITy) process. *Composites Part A: Applied Science and Manufacturing, 40*(8), 1065–1070.
49. Han, C. D., Lee, D. S., & Chin, H. B., (1986). Development of a mathematical model for the pultrusion process. *Polymer Engineering & Science, 26*(6), 393–404.
50. Hashim, M. H. M., Sam, A. R. M., & Hussin, M. W., (2011). The future of external application of fiber-reinforced polymer in civil infrastructure for tropical climates region. *International Journal of Mechanical and Materials Engineering, 6*(2), 147–159.
51. Hinton, M. J. K. A., Soden, P. D., & Kaddour, A. S., (2004). *Failure Criteria in Fiber Reinforced Polymer Composites: The World-Wide Failure Exercise* (Vol. 1, pp. 1–22). Elsevier.
52. Hnat, J. G., (1991). Recycling of insulation fiberglass waste. *Glass Production Technol. Int. Sterling Publications, 1991*, 81–84.
53. Hofstätter, T., Pedersen, D. B., Tosello, G., & Hansen, H. N., (2017). State-of-the-art of fiber-reinforced polymers in additive manufacturing technologies. *Journal of Reinforced Plastics and Composites, 36*(15), 1061–1073.
54. Hou, T. H., & Jensen, B. J., (2004). *Evaluation of Double-Vacuum-Bag Process for Composite Fabrication, 1*, 1–20.
55. Jain, R., & Lee, L., (2012). *Fiber Reinforced Polymer (FRP) Composites for Infrastructure Applications: Focusing on Innovation, Technology Implementation and Sustainability* (Vol. 1, pp. 1–20). Springer Science & Business Media.
56. Jang, B. Z., (1994). *Advanced Polymer Composites: Principles and Applications* (Vol. 1, pp. 1–20). ASM International, Materials Park, OH 44073-0002, USA, 1994. 305.
57. Jassal, M., & Ghosh, S., (2002). *Aramid Fibers: An Overview, 1*, 1–20.
58. Jiao, H., Mashiri, F., & Zhao, X. L., (2012). A comparative study on fatigue behavior of steel beams retrofitted with welding, pultruded

CFRP plates and wet layup CFRP sheets. *Thin-Walled Structures, 59*, 144–152.

59. Joubaud, L., Achim, V., & Trochu, F., (2005). Numerical simulation of resin infusion and reinforcement consolidation under flexible cover. *Polymer Composites, 26*(4), 417–427.
60. Karbhari, V. M., & Abanilla, M. A., (2007). Design factors, reliability, and durability prediction of wet layup carbon/epoxy used in external strengthening. *Composites Part B: Engineering, 38*(1), 10–23.
61. Kim, Y. J., (2019). State of the practice of FRP composites in highway bridges. *Engineering Structures, 179*, 1–8.
62. Klompen, E. T. J., (2005). *Mechanical Properties of Solid Polymers: Constitutive Modeling of Long- and Short-Term Behavior, 1*, 1–23.
63. Kuhn, H. H., Child, A. D., & Kimbrell, W. C., (1995). Toward real applications of conductive polymers. *Synthetic Metals, 71*(1–3), 2139–2142.
64. Kumar, A. A., & Sundaram, R., (2016). Cure cycle optimization for the resin infusion technique using carbon nanotube additives. *Carbon, 96*, 1043–1052.
65. Landesmann, A., Seruti, C. A., & Batista, E. D. M., (2015). Mechanical properties of glass fiber reinforced polymers members for structural applications. *Materials Research, 18*(6), 1372–1383.
66. Lee, C. C., Folgar, F., & Tucker, C. L., (1984). *Simulation of Compression Molding for Fiber-Reinforced Thermosetting Polymers, 1*, 1–22.
67. Lefever, S., Dal, M., & Matthiasdottir, A., (2007). Online data collection in academic research: Advantages and limitations. *British Journal of Educational Technology, 38*(4), 574–582.
68. Lindhagen, J., & Berglund, L., (1998). Microscopical damage mechanisms in glass fiber reinforced polypropylene. *Journal of Applied Polymer Science, 69*(7), 1319–1327.
69. Liu, D., & Xia, F., (2010). Assessing object-based classification: Advantages and limitations. *Remote Sensing Letters, 1*(4), 187–194.
70. Liu, W., Zhang, X., Xu, G., Bradford, P. D., Wang, X., Zhao, H., & Zhu, Y., (2011). Producing superior composites by winding carbon nanotubes onto a mandrel under a poly(vinyl alcohol) spray. *Carbon, 49*(14), 4786–4791.

71. Liu, X. L., Crouch, I. G., & Lam, Y. C., (2000). Simulation of heat transfer and cure in pultrusion with a general-purpose finite element package. *Composites Science and Technology, 60*(6), 857–864.

72. Lötters, J. C., Olthuis, W., Veltink, P. H., & Bergveld, P., (1997). The mechanical properties of the rubber elastic polymer polydimethylsiloxane for sensor applications. *Journal of Micromechanics and Microengineering, 7*(3), 145.

73. Lu, X., Xu, G., Hofstra, P. G., & Bajcar, R. C., (1998). Moisture-absorption, dielectric relaxation, and thermal conductivity studies of polymer composites. *Journal of Polymer Science Part B: Polymer Physics, 36*(13), 2259–2265.

74. Masuelli, M. A., (2013). Introduction of fiber-reinforced polymers-polymers and composites: Concepts, properties, and processes. In: *Fiber Reinforced Polymers-the Technology Applied for Concrete Repair* (Vol. 1, pp. 1–20). IntechOpen.

75. Meijer, H. E., & Govaert, L. E., (2005). Mechanical performance of polymer systems: The relation between structure and properties. *Progress in Polymer Science, 30*(8, 9), 915–938.

76. Mertiny, P., & Ellyin, F., (2002). Influence of the filament winding tension on physical and mechanical properties of reinforced composites. *Composites Part A: Applied Science and Manufacturing, 33*(12), 1615–1622.

77. Miller, A. H., Dodds, N., Hale, J. M., & Gibson, A. G., (1998). High speed pultrusion of thermoplastic matrix composites. *Composites Part A: Applied Science and Manufacturing, 29*(7), 773–782.

78. Miller, M. B., & Tang, Y. W., (2009). Basic concepts of microarrays and potential applications in clinical microbiology. *Clinical Microbiology Reviews, 22*(4), 611–633.

79. Miyagi, K., & Teramoto, Y., (2019). Function extension of dual-mechanochromism of acylated hydroxypropyl cellulose/synthetic polymer composites achieved by "moderate" compatibility as well as hydrogen bonding. *Polymer, 174*, 150–158.

80. Morshed, S. A., Sinha, A., Zhang, Q., & Tatar, J., (2019). Hygrothermal conditioning of wet-layup CFRP-concrete adhesive joints modified with silane coupling agent and core-shell rubber nanoparticles. *Construction and Building Materials, 227*, 116531.

81. Narula, C. K., Allison, J. E., Bauer, D. R., & Gandhi, H. S., (1996). Materials chemistry issues related to advanced materials applications in the automotive industry. *Chemistry of Materials, 8*(5), 984–1003.
82. Niggemann, C., Song, Y. S., Gillespie, J. W., & Heider, D., (2008). Experimental investigation of the controlled atmospheric pressure resin infusion (CAPRI) process. *Journal of Composite Materials, 42*(11), 1049–1061.
83. Nowicki, M., Richter, A., Wolf, B., & Kaczmarek, H., (2003). Nanoscale mechanical properties of polymers irradiated by UV. *Polymer, 44*(21), 6599–6606.
84. Odian, G., (1991). *Principles of Polymerization* (3rd edn., Vol. 1, pp. 1–20). John Wiley & Sons. Inc.: New York.
85. Ohama, Y., (1997). Recent progress in concrete-polymer composites. *Advanced Cement-Based Materials, 5*(2), 31–40.
86. Park, C. H., Lebel, A., Saouab, A., Bréard, J., & Lee, W. I., (2011). Modeling and simulation of voids and saturation in liquid composite molding processes. *Composites Part A: Applied Science and Manufacturing, 42*(6), 658–668.
87. Park, S. Y., Choi, C. H., Choi, W. J., & Hwang, S. S., (2019). A comparison of the properties of carbon fiber epoxy composites produced by non-autoclave with vacuum bag only prepreg and autoclave process. *Applied Composite Materials, 26*(1), 187–204.
88. Partap, S., Hebb, A. K., Rehman, I. U., & Darr, J. A., (2007). Formation of porous natural-synthetic polymer composites using emulsion templating and supercritical fluid assisted impregnation. *Polymer Bulletin, 58*(5), 849–860.
89. Pearson, T. A., (2002). New tools for coronary risk assessment: What are their advantages and limitations? *Circulation, 105*(7), 886–892.
90. Peck, R. B., (1969). Advantages and limitations of the observational method in applied soil mechanics. *Geotechnique, 19*(2), 171–187.
91. Peters, S. T., Humphrey, W. D., & Foral, R. F., (1991). *Filament Winding-Composite Structure Fabrication, 1*, 1–22.
92. Pillai, K. M., (2004). Modeling the unsaturated flow in liquid composite molding processes: A review and some thoughts. *Journal of Composite Materials, 38*(23), 2097–2118.
93. Ploeg, A. V. D., & Claessen, K., (2015). Practical principled FRP: Forget the past, change the future, FRPNow!. In: *Proceedings of the 20th ACM*

SIGPLAN International Conference on Functional Programming (Vol. 1, pp. 302–314).

94. Popescu, D., Zapciu, A., Amza, C., Baciu, F., & Marinescu, R., (2018). FDM process parameters influence over the mechanical properties of polymer specimens: A review. *Polymer Testing, 69*, 157–166.
95. Prokhorov, A. M., (1970). *Great Soviet Encyclopedia, 1*, 1–20.
96. Rana, R. S., & Purohit, R., (2017). A Review on mechanical property of sisal glass fiber reinforced polymer composites. *Materials Today: Proceedings, 4*(2), 3466–3476.
97. Rogovina, S. Z., (2016). Biodegradable polymer composites based on synthetic and natural polymers of various classes. *Polymer Science Series C, 58*(1), 62–73.
98. Rosato, D. V., & Rosato, D. V., (2004). *Reinforced Plastics Handbook, 1*, 1–28. Elsevier.
99. Roy, S., Bhavan, V. C., & Lake, S., (2019). *A Holistic Discussion on Biodiversity, 1*, 1–20.
100. Ruoff, R. S., Qian, D., & Liu, W. K., (2003). Mechanical properties of carbon nanotubes: Theoretical predictions and experimental measurements. *Comptes Rendus Physique, 4*(9), 993–1008.
101. Rusnakova, S., Capka, A., Fojtl, L., Zaludek, M., & Rusnak, V., (2016). Technology and mold design for production of hollow carbon composite parts. *Manufacturing Technology, 16*(4), 799–804.
102. Schell, J. S. U., Deleglise, M., Binetruy, C., Krawczak, P., & Ermanni, P., (2007). Numerical prediction and experimental characterization of mesoscale-voids in liquid composite molding. *Composites Part A: Applied Science and Manufacturing, 38*(12), 2460–2470.
103. Schillfahrt, C., Fauster, E., & Schledjewski, R., (2017). Influence of process pressures on filling behavior of tubular fabrics in bladder-assisted resin transfer molding. *Advanced Manufacturing: Polymer & Composites Science, 3*(4), 148–158.
104. Schilling, C. H., Schuster, S., Palsson, B. O., & Heinrich, R., (1999). Metabolic pathway analysis: Basic concepts and scientific applications in the post-genomic era. *Biotechnology Progress, 15*(3), 296–303.
105. Seitz, J. T., (1993). The estimation of mechanical properties of polymers from molecular structure. *Journal of Applied Polymer Science, 49*(8), 1331–1351.

106. Shi, C., (2009). Corrosion of glasses and expansion mechanism of concrete containing waste glasses as aggregates. *Journal of Materials in Civil Engineering, 21*(10), 529–534.

107. Šimáček, P., & Advani, S. G., (2004). Desirable features in-mold filling simulations for liquid composite molding processes. *Polymer Composites, 25*(4), 355–367.

108. Smallman, R. E., & Bishop, R. J., (1999). *Modern Physical Metallurgy and Materials Engineering* (Vol. 1, pp. 1–20). Butterworth-Heinemann.

109. Stringer, L. G., (1989). Optimization of the wet lay-up/vacuum bag process for the fabrication of carbon fiber epoxy composites with high fiber fraction and low void content. *Composites, 20*(5), 441–452.

110. Suzuki, M., (1994). Activated carbon fiber: Fundamentals and applications. *Carbon, 32*(4), 577–586.

111. Thakur, V. K., Thakur, M. K., Raghavan, P., & Kessler, M. R., (2014). Progress in green polymer composites from lignin for multifunctional applications: A review. *ACS Sustainable Chemistry & Engineering, 2*(5), 1072–1092.

112. Thompson, A. C., Cadusch, P. J., Robertson, D. F., Stoddart, P. R., & Wade, S. A., (2012). Origins of spectral changes in fiber Bragg gratings due to macro bending. *Journal of Light Wave Technology, 30*(22), 3500–3511.

113. Tong, L., Mouritz, A. P., & Bannister, M. K., (2002). *3D Fiber Reinforced Polymer Composites, 1*, 1–20. Elsevier.

114. Tooley, F. V., (1991). Fiberglass. *ASM International, Engineered Materials Handbook, 4*, 402–408.

115. Trochu, F., Ruiz, E., Achim, V., & Soukane, S., (2006). Advanced numerical simulation of liquid composite molding for process analysis and optimization. *Composites Part A: Applied Science and Manufacturing, 37*(6), 890–902.

116. Van, D. V. K., & Kiekens, P., (2001). Thermoplastic pultrusion of natural fiber-reinforced composites. *Composite Structures, 54*(2, 3), 355–360.

117. Wang, Y., Serrano, S., & Santiago-Avilés, J. J., (2003). Raman characterization of carbon nanofibers prepared using electrospinning. *Synthetic Metals, 138*(3), 423–427.

118. Webb, R. O., (2009). Major forces impacting the fiberglass insulation industry in the 1990s. In: *51st Conference on Glass Problems* (Vol. 136, p. 426). John Wiley & Sons.
119. Wegian, F. M., & Abdalla, H. A., (2005). Shear capacity of concrete beams reinforced with fiber-reinforced polymers. *Composite Structures, 71*(1), 130–138.
120. Weissenböck, J., Amirkhanov, A., Li, W., Reh, A., Amirkhanov, A., Gröller, E., & Heinzl, C., (2014). Fiber scout: An interactive tool for exploring and analyzing fiber-reinforced polymers. In: *2014 IEEE Pacific Visualization Symposium* (Vol. 1, pp. 153–160). IEEE.
121. Whybark, D. C., & Williams, J. G., (1976). Material requirements planning under uncertainty. *Decision Sciences, 7*(4), 595–606.
122. Williams, C., Summerscales, J., & Grove, S., (1996). Resin infusion under flexible tooling (RIFT): A review. *Composites Part A: Applied Science and Manufacturing, 27*(7), 517–524.
123. Wu, Z., Wang, X., & Iwashita, K., (2007). State-of-the-art of advanced FRP applications in civil infrastructure in Japan. *Composites and Polycon, 37*, 1–17.
124. Yi, A. Y., & Jain, A., (2005). Compression molding of aspherical glass lenses-a combined experimental and numerical analysis. *Journal of the American Ceramic Society, 88*(3), 579–586.
125. Zerbst, U., Bruno, G., Buffiere, J. Y., Wegener, T., Niendorf, T., Wu, T., & Schnabel, K., (2021). Damage tolerant design of additively manufactured metallic components subjected to cyclic loading: State of the art and challenges. *Progress in Materials Science, 1*, 1–20. In-press.
126. Zhang, W., Dehghani-Sanij, A. A., & Blackburn, R. S., (2007). Carbon based conductive polymer composites. *Journal of Materials Science, 42*(10), 3408–3418.
127. Zhao, Z., & Gou, J., (2009). Improved fire retardancy of thermoset composites modified with carbon nanofibers. *Science and Technology of Advanced Materials, 1, 1–22.*
128. Zhu, Z., Xu, L., Chen, G., & Li, Y., (2010). Optimization on tribological properties of aramid fiber and $CaSO_4$ whisker reinforced non-metallic friction material with analytic hierarchy process and preference ranking organization method for enrichment evaluations. *Materials & Design, 31*(1), 551–555.

129. Zulfiqar, F., Navarro, M., Ashraf, M., Akram, N. A., & Munné-Bosch, S., (2019). Nano fertilizer use for sustainable agriculture: Advantages and limitations. *Plant Science, 289*, 110270.

Chapter 4 Glass Fiber-Reinforced Polymers

CONTENTS

4.1. Introduction.. 126

4.2. Chemical Composition and Classification.. 127

4.3. Fabrication of the Glass Fibers.. 130

4.4. Forms of the Glass Fibers.. 132

4.5. Properties of Glass Fiber... 135

4.6. Glass Fibers in the Polymer Composites... 139

4.7. Applications.. 144

References.. 150

4.1. INTRODUCTION

The fruitful history of fiber-strengthened composites as very lightweight materials in several engineering structures in previous eras was fundamentally started by the bulk production of GFRP (glass-fiber-reinforced plastics) and glass fibers. Besides the presence of some other reinforcement fibers created from aramid or carbon, the huge majority of fiber-reinforced composites present today are prepared from glass fibers. The reason for this situation lies in the comparatively less cost of glass fibers as compared to the other fibers (Landesmann et al., 2015). Conversely, many other particular characteristics such as high chemical resistance, high tensile strength, or electrical insulation make these glass fibers the perfect reinforcement for numerous applications, which are discussed in this chapter. Even though having been in utilization for nearly 70 years, the composites of glass fiber are very far away from getting obsolete, particularly as today's necessity for saving energy pushes the utilization of lightweight materials in almost all of the transportation industries and particularly in the automotive manufacturing (Figure 4.1) (Adam et al., 2015; Shuaib et al., 2015).

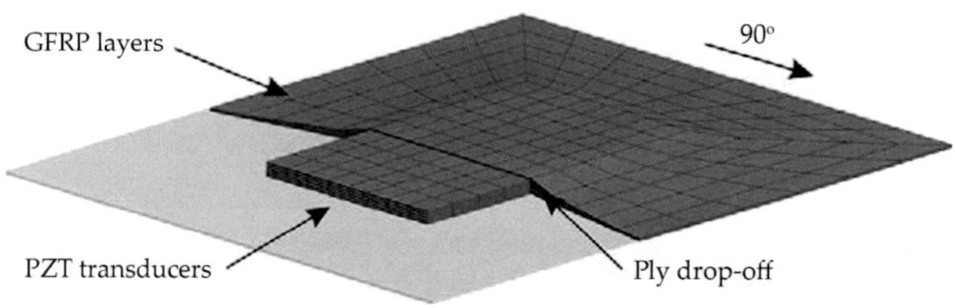

Figure 4.1. *Piezoelectric elements incorporated into the module containing glass fiber reinforced plastics (GFRP).*

Source: https://www.researchgate.net/figure/Piezoelectric-elements-integrated-into-a-module-GFRP-glass-fiber-reinforced-plastics_fig3_323958857.

Even though it is supposed that the glass fibers are 1st reinforcement fibers made by man that has been produced in Roman and Egyptian times, the industrial bulk production of these fibers' dates back to the 1930s. Two revolutionary American companies combined in 1938 and established the Owens-Corning Fiberglas Corporation. This corporation uninterruptedly produced glass filaments for the first time. The innovation as a strengthening fiber for the composite materials then occurred in the 1940s in World War

II consequential from electronic needs. The military aircrafts' radomes were made from GFRP, which marks the first significant use of this material. Generally, most of the historical advancements in the area of glass fiber composites and glass fibers go back to military and aeronautic applications. Since the beginning of the fiber-strengthened plastics industry in those days, GFRP discovered its way into various markets and applications, for instance, in aerospace, marine, automotive, wind energy, leisure products, and construction (Palanikumar et al., 2006; Shuaib et al., 2015). This chapter provides a summary of the basics and recent developments of glass fiber-strengthened polymers in terms of the chemical composition and diverse types, properties, applications, and fabrication.

4.2. CHEMICAL COMPOSITION AND CLASSIFICATION

4.2.1. Chemical Structure of the Glass

Several types of glass fiber are accessible for technical uses, each of them having a different chemical composition triggering exclusive properties for special objectives. All of those kids have in familiar that they are centered on silica, SiO_2, which prevails in pure form as the polymer, $(SiO_2)_n$. Silica possesses no definite melting point, but at temperatures nearly 2000°C, it unstiffens and the molecules are free to move. If the silica is cooled very rapidly from such high temperatures, the molecules cannot form the crystalline structure as in the case of quartz. Even though the 3-D network is established, the glass's structure is amorphous (non-crystalline) with generally isotropic properties and without orientation (Figure 4.2). The high operating temperature of the silica is a major drawback. Thus, diverse other molecules, generally oxides, are familiarized into the glass to alter the network structure and lower its operating temperature, and also to give other features for special applications. Consequential from this selection of surplus constituents, several different kinds of glass fibers have normally been made during the previous decades, which vary in the percentage (%) of the surplus molecules and the consequential global properties. These kinds, which are normally abbreviated with the individual letter, are introduced shortly as follows (Alsayed et al., 1999; Bagheri et al., 2019).

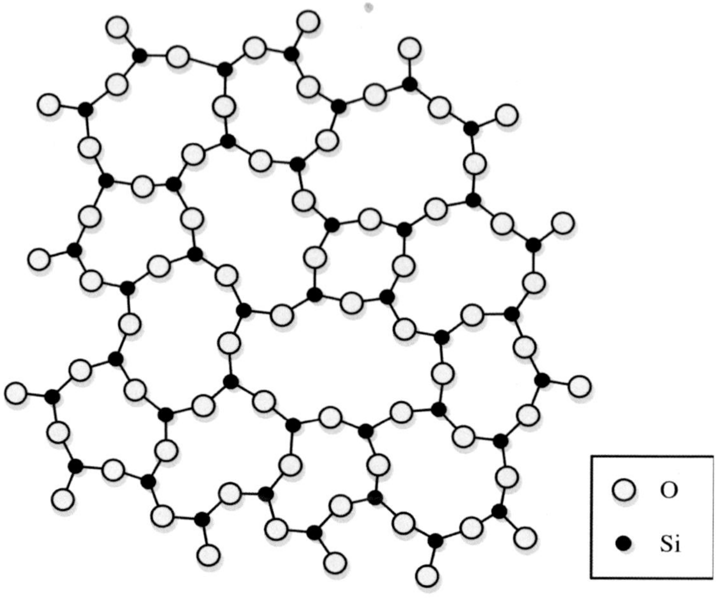

Figure 4.2. Representation of glassy silica's chemical network.

Source: https://www.chegg.com/homework-help/questions-and-answers/different-effects-impurities-fluxes-silicate-ceramic-characteristics-network-modifiers-oxi-q32194915.

4.2.2. Types of Glass Fiber

- **Alkali-Glass:** The first alternative for the mass production of glass fiber was or alkali-glass or alkali-lime glass, which is found in bottles or windows. This glass is inexpensive but exhibits poor resistance to alkali. Today it is normally not utilized for glass fibers any longer (Kalali and Kabir, 2012).
- **Electrical-Glass:** It was established as the 2nd form of glass fiber, termed after its radio-signal transparency and electrical insulation properties. Due to these reasons, the alumino borosilicate glass was an excellent fiber for its first applications in the radomes of aircraft, where transparency of the electromagnetic signals is very important. This was the first type to be developed in filaments and till now forms almost 90% of all the produced glass fibers globally for the glass fiber-strengthened composites (Bunsell, 1988).

- **Chemical-Glass:** This having a high content of boron oxide gives an upsurged corrosion resistance to the acids as compared to electrical-glass, which might be destroyed timely in such kind of environment. Thus, it is commonly utilized for chemical applications. Dielectric-glass: Dielectric-glass is the borosilicate glass that possesses a low dielectric constant and was developed especially for radomes of modern aircraft and covers of the radar antennas (Moustafa and ElGawady, 2016).
- **Low Loss-Glass:** This is a currently familiarized glass fiber by AGY exhibiting the low dissipation factor and low dielectric constant suited ideally for the printed wiring board uses and designs needing increased speed of signal and improved signal integrity as compared to the traditional electrical-glass.
- **Modulus-Glass:** YM-31A or Modulus-glass is the high modulus glass fiber, exhibiting higher mechanical rigorousness properties as compared to electrical-glass. It was developed originally in the 1960s under the Air Force contract to enhance the mechanical capability of the glass fiber composites (Lowrie, 1967). It is not produced anymore because of the poisonousness of BeO.
- **Resistance-Glass, Strength-Glass:** These are aluminosilicate glass fibers having higher elastic modulus, tensile strength, fatigue properties, and impact resistance as compared to electrical-glass with strength-glass being the American version and resistance-glass the European version. These kinds of fibers are more resistant to heat as their strength decreases at 250°C and not at 200°C. The advancement dates back to around the 1960s for military applications such as missile casings and aircraft components. Out of all the reinforcement fibers, these fibers possess the highest tensile strength, even higher than aramid fibers and carbon. The dissimilarity in higher manufacturing cost and chemical composition makes them more costly as compared to the conventional electrical-glass fibers. Thus, the cheap version of strength-glass, also known as S2-glass, was made. The major difference is that this glass is made with less strict non-military specifications. Conversely, the elastic modulus and tensile strength are normally similar to strength-glass (Mallick, 2008; Fazilat et al., 2012).
- **Thermal-Glass:** Out of all the common glass fibers this glass

possesses the lowest thermal expansion coefficient. Thermal-glass is the high-performance fiber familiarized by Nittobo Boseki Corporation Limited and a similar Japanese alternative of the American S-glass and European R-glass (Hazen and Dawson, 2007).

Besides the 40% lower CTE (coefficient of thermal expansion) as compared to electrical-glass, it has a 16% higher modulus and 36% higher tensile strength (Middleton, 1990).

- **Zirconia-Glass:** This glass possesses high resistance to alkaline environments, analogous to acid resistance of the chemical-glass, both glasses being made for chemical uses (Bunsell, 2005).

4.3. FABRICATION OF THE GLASS FIBERS

4.3.1. Production of Fiber

Glass fibers are generally produced by the viscous drawing procedure, as demonstrated in Figure 4.3. The furnace, which is loaded continuously with the raw materials, yields molten glass at a temperature of nearly 1400°C to 1700°C. The glass melt is monitored constantly and refined in order to ensure the proper composition. At the viscosity level between approximately 60–100 Pascal-second, this glass flows into the electrically heated Pt-Rh (platinum-rhodium) alloy bushing comprising numerous holes (200 to 400) in its base. Current use of alloys comprising traces of yttria, thoria, or zirconia enables the production of bushing plates having nearly 6000 holes (Table 4.1) (Jones, 2001; Bunsell and Renard, 2005).

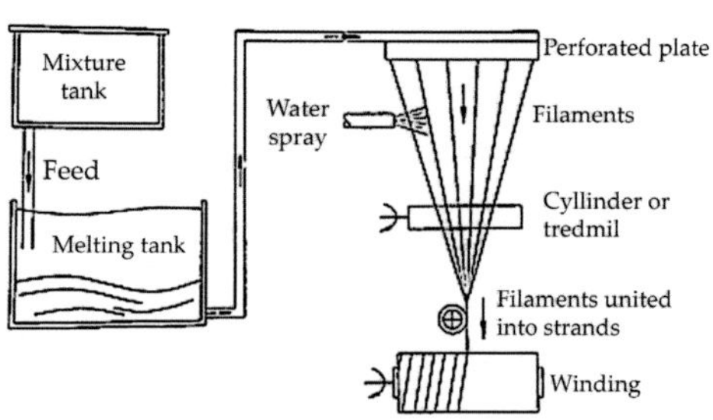

Figure 4.3. Diagram of the production process of glass fiber.

Source: https://www.researchgate.net/figure/6-Schematic-representation-of-a-glass-fiber-production-line-Adapted-from-1_fig3_315690740.

Table 4.1. Chemical Composition of Different Types of Glass Fiber in Weight Percent

Chemical Component	A	C	E	D	M	R	S	Z
Al_2O_3	1.5	4	14	0.2	–	25	25	0.7
SiO_2	72	64	54	74	53.7	60	65	68
MgO	2.3	3	3	0.1	9	6	10	–
CaO	9.5	14.5	19	0.5	12.9	9	–	–
K_2O	0.5	0.5	0.5	1	–	–	–	0.5
Na_2O	13.5	9	0.4	1	–	–	–	11.8
TiO_2	–	–	0.2	0.1	7.9	–	–	1.5
B_2O_3	0.2	5	8.5	23	–	–	–	–
SO_3	0.3	–	–	0.1	–	–	–	–
Fe_2O_3	0.2	–	0.2	–	0.5	–	–	–
ZrO_2	–	–	–	–	2	–	–	16.5
Li_2O	–	–	–	–	3	–	–	1
BeO	–	–	–	–	8	–	–	–
CeO_2	–	–	–	–	3	–	–	–

The glass drops evolving from every hole are usually drawn into just 1 filament by pulling at the speed of 50 ms^{-1}. The diameter of subsequent filament, normally around 5 to 20 mm, is dependent on the bushing hole's size, the viscosity (Pa-s) of glass melt, and the pulling speed (ms^{-1}) of the filament (Baker et al., 2004).

Because of the substantial diameter reduction between the filament and the bushing hole a quick cooling rate is accomplished (>10,000 C/s). This quick cooling is needed to freeze the inner molecular structure between 200°C and 300°C, which outcomes in considerably higher chemical resistance and modulus compared to the bulk glass (Baker et al., 2004). These filaments are cooled further by the sprayed water.

4.3.2. Sizing Application

The next phase in the glass fiber production is the veneer of filament with the sizing. The sizing agent is normally applied as an emulsion or aqueous dispersion in combination with the rubber roller. This is a critical component for the glass fibers and comprises various components. The basic component is the lubricant, frequently the mineral oil, which acts as an agent of protection against the highly coarse nature between the aids and individual filaments in further steps of processing like filament weaving or winding (Deák and Czigány, 2009). The binder component is also included in sizing to hold or bond the filaments together into the glass fiber strand. Usual polymeric binders are polyacrylate, polystyrene (PS), or polyvinyl acetate. The third main component in the sizing is a coupling agent that serves as the adhesion promoter amongst the resin and the filaments in a later composite. Typical materials utilized as coupling agents are polyvinyl acetate, silane, or polymers based on polyester, polyurethane (PUR), or epoxy (Soares et al., 2015).

After the sizing agent application, a filaments bundle is gathered into the strand that normally contains 204,102, or 52 fibers, dependent on the number of the holes in a bushing. Lastly, the strands of glass fiber are collected on the spool and are conveyed to the drying oven, where the moisture is eliminated and the sizing components are cured (Agarwal et al., 2006).

4.4. FORMS OF THE GLASS FIBERS

4.4.1. Commercially Accessible Forms of the Glass Fibers

Glass fibers are available commercially in diverse forms, essential for the individual making procedures of glass fiber reinforced plastics composite parts. Many terms are applied for the various forms which are all standardized. The source is the single filaments that are joined to strands during the process of manufacturing with disentangled, parallel orientation. The group of disentangled, parallel strands-frequently 30 to 60-can be joined to form the roving, which is normally wound on the cylindrical spool (Figure 4.4(a)). Such kinds of rovings are utilized for filament winding or pultrusion manufacturing procedures of the composite parts (Li et al., 2014; Małek et al., 2021). Moreover, they are utilized for woven fabrics and the manufacturing of prepreg by infusing fibers with the polymeric matrix. In comparison to the disentangled roving, one or various strands can be warped to form the yarn, with normally 28 to 40 twists m^{-1}. By this approach, both workability and integrity are augmented for processing the textile fabrics. The woven fabrics with bidirectional properties are possibly the most adaptable form that can generally be utilized from hand lay-up to autoclave processing (Sims and Broughton, 2000) (Figure 4.4(b)). The properties are dependent on the pattern of weave (e.g., twill weave, plain weave, satin weave), yarn type, and the density of a yarn. In addition to 2-D woven fabrics, interweaved textile preforms are used increasingly in engineering applications because of high flexibility in the complex geometries, probable reinforcement angles, and variations in the wall thickness and the cross-section shape. Figure 4.4(c) exhibits the glass fibers tubular braid as a work for the composite tube (Gessler et al., 2007).

Chopped strands are also a form of glass fibers, which are made by cutting strands or rovings into short lengths, normally 50, 25, 12, 6, 3, 1.5 mm. These shorter lengths are utilized for the operations of injection-molding, whereas longer strands are utilized to form CSM (chopped strand mats). Such non-woven mats are made by compressing the chopped strand with arbitrary orientation on the flat surface utilizing a chemical binder, triggering almost equivalent in-plane properties in all of the directions (Figure 4.4(d)). Their usual application is hand lay-up moldings. The continuous strand mats are available, created from the unchopped strand that is interlocked mechanically and needs less chemical binder, occasioning in improved draping qualities appropriate for the matched-die molding (Havar and Stuible, 2009).

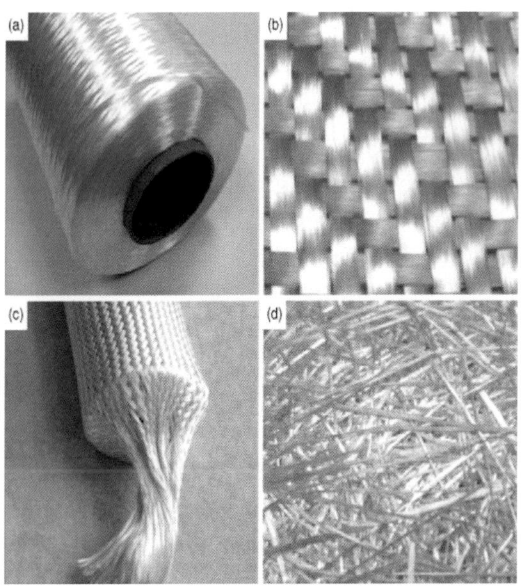

Figure 4.4. *Images of diverse forms of the glass fibers: (a) roving on a coil; (b) woven fabric; (c) braided fabric; (d) chopped strand mat.*

Source: https://onlinelibrary.wiley.com/doi/abs/10.1002/9783527645213.ch6.

The continuous glass fiber or short glass fiber mats are used additionally for GMT (glass mat thermoplastic) composites, for instance, for industrial production of thermoplastic automobile parts (Drechsler et al., 2008).

The chopped strands in amalgamation with thermosetting or thermoplastic resins are well used for the molding compounds. Usual examples based on the thermoset matrix are SMC (sheet molding compound) or BMC (bulk molding compound), which is, for instance, utilized for high-volume cheap automotive parts. One more type is the milled fibers having the typical length of nearly 0.08 to 6 mm, developed by hammer milling continuous strands or chopped strands (Li et al., 2004).

4.4.2. Shaped Glass Fibers

In addition to the circular glass fibers defined so far, broad research on the shaped glass fibers was carried out that are categorized by the cross-sectional shape. Such studies occasioned early requirements for composite materials having higher ratios of stiffness to weight. Also, numerous natural fibers possess shapes other than the regular cylinders. Ideally, glass fibers can generally be drawn in the diversity of diverse shapes by choosing a suitable nozzle geometry in the process of manufacturing (Eckerskorn et al., 1988; Hucker et al., 2003).

One usual form of the shaped glass fibers is a hollow fiber having a circular cross-section, which might offer a 30% lower density with 1.8 g cm^{-3} compared to traditional glass fibers with 2.5 g cm^{-3}. Significantly, the hollow interior of fiber is not filled with the resin, else the weight advantage would be lost. The key potential of hollow fiber is the augmented bending stiffness specific to weight and buckling strength under the compression loads because of an upsurge in the moment of inertia associated with the solid circular fiber of the same mass/length. The resistance of fibers to kink band formation or micro-buckling upsurges, which is the controlling failure beginning mode under the compressive loads on a microscale (Heimbs et al., 2010; Ding et al., 2019). A substitute to the hollow round fiber is the hollow hexagonal fiber.

Moreover, rod-shaped fibers having hexagonal, triangular, or square cross-sections have generally been examined (Humphrey, 1967). Appropriately shaped fibers can usually be packed together efficiently, occasioning in the composite materials having a higher volume fraction of fiber and thus higher strength and weight-specific stiffness properties. The contrast of the tensile strength of single fiber with triangular and circular geometry of analogous cross-sectional area displayed a 25% higher strength for a triangular shape and the compressive strength is 60% higher. This drift was also confirmed for the corresponding composite materials having analogous volume content of fiber, where the compressive strength was 40% higher and the tensile strength was 20% higher with the triangular fibers. This was accredited to the better fiber arrangement inside the composite and an upsurge in the moment of inertia (Karakaya et al., 2020; Bond et al., 2002).

4.5. PROPERTIES OF GLASS FIBER

4.5.1. General Properties

The glass density is mainly dependent on the glass composition, hence for diverse kinds of glass, a marginally different density is discovered. The density of fiber is generally marginally lower as compared to the value for bulk material (2.49 to 2.52 g cm^{-3} for strength-glass, 2.54 to 2.58 g cm^{-3} for electrical-glass). Generally, the refraction index, which is concluded for light with the wavelength of 550 nanometers is directly proportional to glass density and also changes depending on the composition of glass. For all types of glass utilizes for the production of fiber the refraction index between 1.45 and 1.65 is measured (Praeger et al., 2012; Bhoopathi et al., 2014).

4.5.2. Elastic Properties

The classification of Young's modulus for the glass fibers is not straightforward because of the trouble in evaluating the elastic properties of tiny fibers. Various experimental examinations have established that there is no or little elastic anisotropy in the glass fibers, which considerably decreases the experimental struggle to determine the elastic properties. Young's modulus is concluded either by acoustic or mechanical tests. Mechanical test approaches are frequently tensile or bending tests, which need as input a precise fiber diameter, but also permit for the instantaneous measurement of elongation and strength at failure (Samal et al., 2009; Nam et al., 2017). The acoustic approach concludes the velocity of the elastic wave traveling within the fiber. The acoustic approach can be applied at elevated temperatures very easily. The Poisson's ratio and shear modulus of fiber is determined by the supplementary torsion-resonance experiment, where the torsion pendulum is adjourned on a fiber itself. Usual values of Young's modulus for the glass fiber is between 70 and 75 GPa for alkali, chemical-, and electrical-glass while higher values are attained for high strength (strength- and resistance-) and high modulus (modulus-) glass fibers, 85–115 GPa, correspondingly. The values given are effective at room temperature and generally slightly decrease at higher temperatures. This dissimilarity is more noticeable for strength-, resistance-, and modulus-glasses as for electrical-glass. These values of stiffness are considerably lower in contrast to the other reinforcing fibers such as aramid or carbon. Because of this dissimilarity, glass fibers and composites are not appropriate for ultra-lightweight structures (Puska et al., 2004; Dikbas et al., 2013).

4.5.3. Elongation and Strength Properties

The mechanical and physical properties of bulk glass and glass fiber are frequently very comparable apart from strength. The considerably higher rate of cooling during the production procedure of glass fiber associated with the bulk glass is a basis of this dissimilarity. The strength of the virgin glass fibers is considerably higher as compared to the strength of fibers confirmed at the end of the production process. This dissimilarity is triggered by the consequent steps since the glass fiber is sensitive to any additional mechanical contact, for example during sizing application or in the process of winding (Park and Jin, 2001). Usual final values of strength between 2.3 and 2.8 GPa can be discovered for alkali-, chemical-, electrical-, and modulus-glass, while high strength (resistance- and strength-) glass fibers have the tensile strength between 3.1 and 3.9 GPa (Table 4.2).

Table 4.2. Properties of the Commercial Glass Fibers

Property	**Unit**	**Glass Fiber Type**						
		A	**C**	**D**	**E**	**M**	**R**	**S**
General properties Refraction index	–	1.512	1.541	1.47	1.548	1.635	–	1.523
Filament diameter	mm	–	–	–	5–20	–	–	8–14
Density	g/cm^3	2.47	2.49	2.16	2.54	2.89	2.49	2.49

Elastic properties Shear modulus	GPa	–	–	–	29.6	–	–	36
Tensile modulus	GPa	73	74	55	71	115	86	85
Poisson's ratio	–	–	–	–	0.20	–	–	0.22
Tensile strength (roving)	GPa	2.76	2.35	2.41	2.4	2.4	3.1	3.91
Tensile strength (Virgin filament)	GPa	3.1	3.1	2.5	3.4	3.4	4.4	4.58
Chemical resistance Weight loss after 1 h in boiling water	%	11.1	0.13	–	1.7	–	–	–
Stain at failure	%	3.6	3.5	–	2–4	–	4.2	5.4
Weight loss after 1 h in boiling NaOH (0.1 N)	%	12–15	2.28	–	9.7	–	–	–
Weight loss after 1 h in boiling H_2SO_4 (1.0 N)	%	6.2	0.10	–	48.2	–	–	–
Thermal properties Specific heat capacity	J/(kgK)	–	836	732	800	–	–	735
Thermal conductivity	W/(Km)	–	–	–	1.1	–	–	–
Maximum use temperature	C	–	–	–	300	–	350	–

CTE	10^6/K	9	7–8	2–3	4.9–6	–	4	2.9–5.6
Loss tangent at 10^6 Hz	10^6	–	–	500	3900	–	1500	7200
Electrical properties Relative dielectric constant	–	6.9	6.2–6.3	3.5–3.8	5.9–6.6	–	6.2	4.5–5.6
Volume resistivity	Vm	10^8	–	–	1013	–	–	–

The high values of strength are certainly one of the exceptional properties of the glass fibers, surpassing other reinforcing fibers such as aramid and HM carbon. The HT carbon fibers can exhibit analogous strength values (Hao et al., 2013).

The glass fibers' tensile strength is impacted by several environmental effects. The first effect is discovered in the preparation of glass melt during the process of manufacturing. The conditioning temperature considerably impacts the ultimate strength and its variation coefficient; hence for every type of glass fiber, optimum processing conditions should be maintained at all times. The glass fibers also undergo corrosion when placed inside humid environments (Laura et al., 2002; Lu et al., 2009).

Frequently the sizing agent of glass fiber will serve as the protection layer and will try to avert the process of corrosion. Another main problem is strain rate dependence of the glass fiber's mechanical properties, which are particularly important for studies or applications, where high rates of loading are involved, for instance, impact loads (Wu et al., 2001).

The glass fibers conditioning at elevated temperatures and for extended times has a decreasing impact on the glass fibers' strength. The strength of the glass fibers is lower when verified at higher temperatures because of the sizing decomposition where the consequential chemicals deteriorate the fiber. The strength of the glass fibers is very delicate to handling and extreme care should be taken to avoid any fiber contact with the other solid at test gage. Also, the diameter of fiber should be determined with utmost care as it has a noticeable impact on the measurement results. The measurement of strength at higher temperatures as compared to the regular room conditions might be carried out by local heating of fiber within the gage length.

4.5.4. Corrosion Properties

As mentioned previously, the glass fibers' corrosion resistance is generally fully governed by the sizing of fiber. Non-coated fibers would considerably corrode in acid or alkali environments or water would also deteriorate the fiber's properties. Specifically developed compositions of glass in chemical-glass and zirconia-glass have the higher corrosion resistance as compared to electrical-glass in alkaline and acidic environments, correspondingly. The fibers corrosion resistance is concluded by the glass fiber's weight loss measurement when boiling in acid or alkali liquids or the distilled water for one hour (Deitz et al., 2003; Mahrez et al., 2005).

4.5.5. Thermal Properties

The glass fibers' specific heat capacity is marginally lower associated with the bulk glass equivalent value and primarily depends on the composition of glass. The value between 700 and 800 J kg^{-1} K^{-1} is available in the literature. The thermal conductivity of the electrical-glass fibers is nearly 1.1 W m^{-1} K^{-1}. The thermal expansion coefficient changes for the diverse compositions of glass and is discovered between 210^6 K^{-1} and 910^6 K^{-1} (Table 4.2).

Glass fibers possess very high electrical and thermal resistance and are therefore often utilized in electrical and electronic components. L- and D-glass, which were particularly made for the high transparency of signal, have a considerably lower dielectric constant as compared to the other glass types (Chan et al., 1978).

4.6. GLASS FIBERS IN THE POLYMER COMPOSITES

4.6.1. Polymers for the Glass Fiber-Strengthened Composites

Glass fibers are utilized as strengthening material in amalgamation with both thermoset and thermoplastic resins. The processes of manufacturing utilized for thermoplastics are centered on melting a polymer by familiarizing heat and then solidifying the blend of polymer-fiber after forming. Because of the high viscosity of the thermoplastic polymer, the matrix material is frequently premixed with a glass fiber reinforcement (Chen et al., 2014; Sathishkumar et al., 2014). For the long glass fibers, this is accomplished in the shape of commingled yarns or the polymer-fabric layered preform, while for the short fiber-strengthened plastics, for instance, the fibers are generally premixed in polymer pellets for injection molding. Particularly in the circumstance of the short fiber-strengthened material, the local alignment of fiber is dependent strongly on the flow of polymer during the process of manufacturing, which can considerably impact the physical properties of composite material. Instances of the thermoplastic materials commonly utilized for the glass fiber-strengthened composites are PA (polyamide), PE (polyethylene), PP (polypropylene), PC (polycarbonate), PS, ABS (acrylonitrile butadiene styrene), POM (polyoxymethylene), PBT (polybutylene terephthalate), and PEEK (polyether ether ketone).

Thermosetting resins are joined with the long glass fibers inside either infiltration procedures (e.g., vacuum-assisted process or resin transfer molding (RTM)) or are accessible as the pre-impregnated material. SMC, BMC, and Spray-up processes are usually applied in amalgamation with the short fibers. After forming, vitrification of a thermoset is begun at elevated temperatures of the one-component resin system or through mixing the resin system's components utilizing two or more components. UP (Polyester), PI (polyimide), VE (vinyl ester), PUR, PF (phenolic), and EP (epoxy) resins are instances of the thermosets that are strengthened with the glass fibers (Wu et al., 1997; Yao et al., 2015).

4.6.2. Determination of Properties

As in all of the composite materials, the physical properties of an occasioning material are reliant on the physical behavior, the concentration, and the form of all the constituents, along with their compatibility. Precisely for the compatibility reason, various coupling agents in fiber sizing are utilized to improve the adhesion between glass fiber and resin, and hence enhancing the composite's properties. The diverse forms of the glass fibers as defined earlier are primarily chosen based on their workability and performance but also cost might be the selection criterion (Mansor et al., 2013; Xu et al., 2013).

To conclude the properties of consequential composite, both analytical and experimental methods can be utilized. For experimental examinations, various testing standards are accessible and must be used to enhance the reproducibility and quality of the attained results. Conversely, the standards are frequently only effective for a certain kind of reinforcement, and extra care must be taken when trying the other strengthening configurations. For example, the interlacing pattern of the woven glass reinforcement might outcome in local discrepancies of mechanical properties, and therefore a broader test specimen is needed than in the situation of a UD (unidirectionally) reinforced composite (Deng et al., 1999a, b). For the analytical determination of physical properties, several models are accessible in literature, which can normally be grouped into three categories. The 1st category is the micro-mechanical formulation, which concludes the physical property as the simple function of base component properties and fiber volume fraction of a composite material. This approach is quite easy in utilization, but only offers inadequate detail and is frequently only precise for a restricted range of volume fractions of a fiber. This method is time economical, easy to utilize as it is executed in diverse software tools but gives information on an averaged behavior and does not give details regarding the field variation within the structure of composite. The 3rd category is centered on multilevel modeling methods and utilizes homogenization methods to conclude the average properties of a reinforced polymer (Kouznetsova et al., 2004). This approach does not only give an average property of a composite but also gives comprehensive information on internal physical phenomena that take place when loaded. The approach of multilevel modeling is very expensive computationally but, particularly for the complex textile-strengthened composites, this approach can normally be the only feasible approach (Xia et al., 1994; Case et al., 1999).

4.6.3. Manufacturing Processes and Associated Composite Properties

Short fiber-strengthened thermoplastic polymers are used commonly to manufacture the injection molded parts because of the augmented mechanical performance, enhanced creep resistance, and toughness compared to the unreinforced parts. The history of flow during injection considerably impacts the orientation of fiber and inclines to form the layer at the surface of a part, where the orientation of fiber considerably diverges for the rest of a part, triggering anisotropic properties. Rate dependency and temperature along with chemical resistance and moisture absorption are primarily governed by a thermoplastic matrix material (Rotem and Lifshitz, 1971; Johnston, 1997).

Short fiber-strengthened composites centered on chopped strands created either by spray or hand lay-up are perhaps the most extensive reinforcement form of the glass fibers because of the low cost of tooling and the process itself. Normally composites having a volume content of fiber between 10% and 25% are manufactured. Elevated volume fractions are hard to accomplish because of the internal reinforcement structure. The large change in quality is the main drawback of the chopped strand composite materials. As a direct outcome, the mechanical capability is strongly dependent on this quality. When designed correctly, the chopped strand composites can generally have better impact performance (Pierard et al., 2004; Li et al., 2014).

One form of the glass fiber-strengthened polymers is discovered in SMC and BMC materials. These materials comprise next to the polymer matrix and the glass fibers a huge amount of the filler material, frequently chalk, to attain low shrinkage features. The arbitrary nature of this reinforcement fiber outcomes in comparatively low physical properties frequently with high variation coefficient.

The profiles of pultruded glass fiber are made in a multiplicity of cross-sections but have a high amount of fibers directed along the direction of the profile, hence having the highly anisotropic behavior. The fragments made with the pultrusion procedure are primarily intended to endure axial tensile loads because of their strength and modulus in this direction. Frequently these parts will not fail due to mechanical failure of material but because of buckling of the Web or flange of the profile consequential from the structural design with thin walls (Sanghera et al., 2009; Landesmann et al., 2015).

The filament winding method, which is frequently utilized for the production of the part in the glass fiber-strengthened thermosetting polymers, gives high-quality parts having high mechanical properties because of the good and governed arrangement of the fibers. The common issue during the winding process of filament is the creating of voids, which normally can impact the performance of material considerably. Usual applications for the glass fiber filament winding are vessels, torque tubes, or pipes (Xia and Wang, 1996; Sierakowski, 1997).

The glass fibers are also utilized as reinforcements in the prepreg tapes, which are normally made by pre-impregnating the fiber with the thermosetting resin. In this type of arrangement, the fibers are normally arranged in the UD flat way and joined with the resin. Such kinds of prepregs need storing at low temperatures (usually 18°C) before utilization in order to avoid cross-linking of a resin. For the production of the composite part, the single plies of prepreg are arranged together in the laminate until the desired thickness is accomplished and cured at higher pressure and temperature. The utmost anisotropic material behavior is accomplished when all of the layers are piled in a similar direction. Laminates of the UD tapes have high strength and in-plane stiffness but lower performance under impact loads and in out of the plane direction. Preimpregnated tapes should be kept at well-governed conditions, occasioning additional expenditures. Usual fiber volume fractions discovered for the pre-impregnated tapes are in the middle of 50–65% (Welsh and Harding, 1998).

4.6.4. Fatigue and Strength Properties

The strength of the glass fiber composites is strongly dependent on the kind of reinforcement and the adhesion quality between fiber and resin. The maximum tensile strength is attained for UD strengthened resin experienced in the direction of reinforcement as the major fraction of load is generally carried out by the glass fibers. For all of the other reinforcement kinds or the off-axis loading, the tensile strength measured will be lower. Because of the point that the GFRP parts have restricted dimensions generally in the direction of thickness, the compressive strength is often hard to measure as the parts buckle easily under the load applied. The shear strength is not straightforward to conclude, although various experimental approaches are available. For the composites having high anisotropic nature, a substantial difference between compression and tensile strengths is found. The difference is less noticeable for the materials having reinforcement in various directions, for instance, braided or woven fabrics, BMC, SMC, or CSM (Heimbs et al., 2007; Witten, 2010).

GFRP possesses outstanding fatigue resistance. For the UD glass fiber-strengthened composite, the consequentially reduced strength after the cycling loading can generally be written as:

$$\sigma = \sigma_U(1 - k\log(N))$$

where; σ_U is the ultimate tensile strength; N is the number of the load cycles; k is the fractional loss factor.

Typical values of K for GFRP at room temperature are in the middle of 0.08–0.1. At elevated temperatures, higher values of k are observed, giving lower fatigue resistance to the material. High strength fibers do not only have higher static strength but higher fatigue strength also as compared to electrical-glass fibers (Jeong et al., 2007).

4.6.5. Effect of Strain Rate in the Glass Fiber-Reinforced Composite

Characteristical for the glass fiber composites is the effect of strain rate, which is primarily accredited to the stiffness depending on the rate and the strength behavior of pure glass fibers, as previously mentioned. Under the high rates of loading in a short period, strength, failure strain, and stiffness of a composite can increase significantly. Failure strain and failure stress can be three or two times higher for the high rates of loading, which must be considered for load cases or applications involving very high strain rates. The effect is more important for the glass fiber composites as compared to the aramid fiber or carbon composites, where such kind of effects only become important at very higher strain rates, and in spite of some upsurge in the strength, the failure strain instead inclines to be not affected or even reduces, respectively (Heimbs et al., 2008).

4.6.6. Environmental Influences

The properties of the glass fiber-strengthened composites are disturbed by temperature, moisture, ultraviolet radiation, and chemical environment. GFRP usually absorbs 0.3 to 1.5% of water, dependent on the temperature, the resin material, the loading conditions, and the exposure time. The water absorption will reduce the mechanical capability of a composite, might affect the dimensions of a part, and can outcome in the chemical weakening of the glass fiber. Temperature changes disturb the physical properties of a part that are highly dependent on the material of resin such as shear and transverse stiffness and strength. At elevated temperatures, the polymer matrix will begin to degrade because of chemical decomposition. Usually, the material will have the maximum operation temperature varying from 90°C for GMT to 200°C for continuously strengthened epoxy resin (Georgiadis et al., 2008).

4.7. APPLICATIONS

Nowadays, in a majority of the fiber-strengthened composites, the strengthening fiber is made up of glass. Thus, the multiplicity of technical uses is multiple. A breakdown of nowadays GFRP production into diverse industrial sectors displays that the area of construction is leading, trailed by the transportation industry, electronic products, leisure, and sports products, and many others. A summary of usual applications is provided here, exhibiting the potential of contemporary glass fiber composites (Van Tooren, 1998).

1. **Civil/Construction Infrastructure:** Several examples of uses of GFRP in the sector of construction exist, for instance, panels, light domes, corrugated panels, lightwells, dormers, roof gutters, door surrounds, chimneys, window canopies, and coping systems. Such kinds of components not just fulfill requirements of durability in

the corrosive media but give benefits in quicker installation and handling because of the decreased weight. Currently, platforms and bridge decks have received much attention due to the high ratio of strength to weight, corrosion resistance, and lightweight (Figure 4.5) (Bull, 2004).

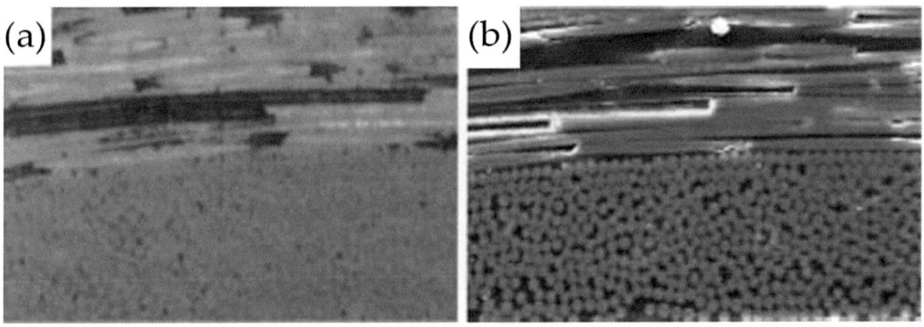

Figure 4.5. Cross-sections of the GFRP sample: (a) optical microscopy; and (b) computed tomography.

Source: https://onlinelibrary.wiley.com/doi/abs/10.1002/9783527645213.ch6.

2. **Aeronautics:** The advancement of GFRP began in the industry of aeronautics and there are many applications today. Even though carbon fibers find their applications in the main structures of commercial airliners, there also exist classical uses for glass fibers, typically in the sandwich structures such as in the entire cabin interior. Exterior uses can be discovered in a radar-signal-translucent radome of an aircraft and the fairing surfaces of empennage and wings. Uses in main airframe structures are restricted because of the comparatively low stiffness. Only in sailplanes and small aircraft the entire fuselage and the wings are frequently made from GFRP. However, glass fibers are used in the upper fuselage shells of the Airbus A380, which is the largest commercial aircraft. These fuselage shells are generally made from glass-reinforced fiber metal laminate (GLARE), which is the layered structure of aluminum sheets and GFRP with exceptional damage tolerance, corrosion, fire resistance, and fatigue properties. The blades of the helicopter are also usually manufactured from GFRP with a honeycomb core or some foam. An outline of the utilization of GFRP in present aeronautic structures is exhibited in Figure 4.6 (Davies et al., 2000).

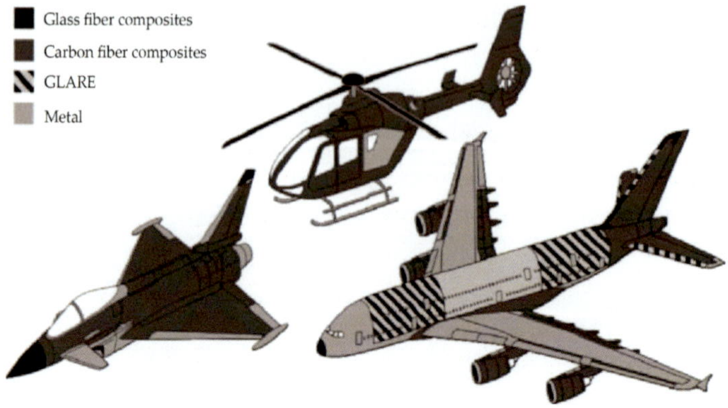

Figure 4.6. Utilization of the composite materials in present aeronautic structures: EC-135 helicopter, Airbus A380 airliner, and Eurofighter jet.

Source: https://onlinelibrary.wiley.com/doi/abs/10.1002/9783527645213.ch6.

3. **Marine:** After the advancement of GFRP for military aircraft, the production of boats (sailing boats, rowing boats, lifeboats, and motorboats,) was the first-ever civilian application, which rapidly gained approval since the 1950s. The manufacturing of boats was an ideal area for both spray lay-up and hand lay-up methods for manufacturing the complex shapes (Figure 4.7).

Figure 4.7. Boat made from GFRP.

Source: http://compositesmanufacturingmagazine.com/2016/11/toyota-unveils-new-powerboat-hybrid-composite-hull/.

Not just the components but the large decks and hulls could also be made this way, modernizing the building of a boat. Nowadays, a growing number of yachts and boats are manufactured from GFRP, in the military and civil sector. One problem of current studies is the influence performance of such kind of structures to projectile impact loads and water slamming (Figure 4.8) (M€akinen, 1995).

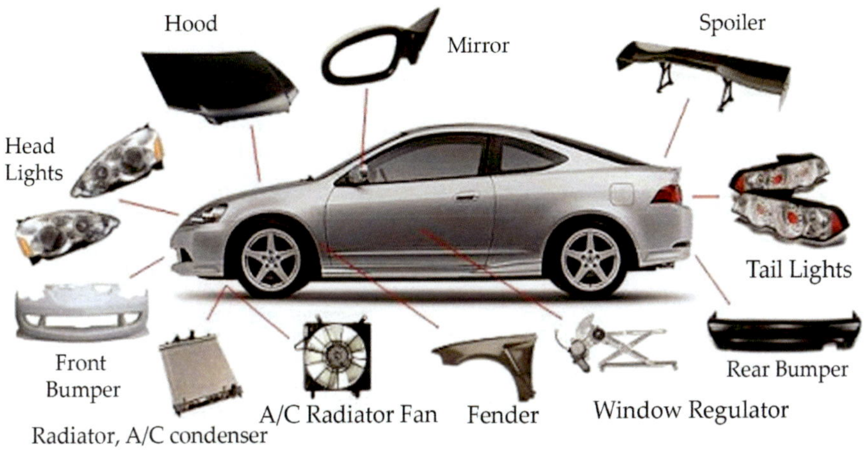

Figure 4.8. Diverse body parts made from GFRP.

Source: http://m.lfrt-plastic.com/news/why-car-front-end-modules-love-long-glass-fibe-11182003.html.

4. **Automotive:** GFRP was 1st utilized in the sequential production of fairings and panels for commercial vehicles (engine hood, wind deflectors). From the 1970s, the bumpers of cars were also manufactured from GFRP. After the institution of SMC, more parts such as trim panels, trunk lids, hoods, and door components were manufactured from such kinds of composite materials. Non-visible parts for the under-bonnet applications like the front bulkhead holding the radiator, front lights, and cooling fan, are also made from GMT because of the lower quality of a surface. The body of small sports cars is made from GFRP. The utilization of composite materials in the automobile industry is considerably increasing in current days following the style of lightweight cars for reduced consumption of energy, and glass fibers are a very good candidate because of their price advantage.

5. **Sporting and Consumer Goods:** Numerous articles in the sporting and leisure industry are generally made from cheap and lightweight glass fiber composites such as tent poles, arrows, pole vault poles, crossbows, and bows, hockey sticks, bowling balls, surfboards, fishing rods, and microwave cookware.
6. **Pipes and Tanks:** Storage tanks, for instance, for the storage of chemicals, are frequently made from GFRP having capacities of nearly 300 tons. Whereas simple and small tanks are manufactured from CSM, high-performance pressure vessels and tanks are usually made by the process of filament winding. The GFRP based pipe systems are utilized for various diverse applications with liquid or the gaseous media of diverse temperatures (Figure 4.9).

Figure 4.9. GFRP based tanks and pipes.

Source: https://www.frpmachine.com/frpmachine/frp-pipe-tank.html.

7. **Wind Energy:** The blades of a wind turbine are classical instances of the large GFRP based structures. This material has been demonstrated to be best appropriate for the mechanical and aerodynamic necessities of the lightweight components such as stiffness, durability, shock resistance, and strength. It is also utilized for the nacelle which is the turbine housing (Figure 4.10).

Figure 4.10. GFRP based blades of wind turbine.

Source: http://compositesmanufacturingmagazine.com/2017/11/global-fiberglass-solutions-finds-success-recycling-turbine-blades/.

REFERENCES

1. Adam, M. A., Said, M., Mahmoud, A. A., & Shanour, A. S., (2015). Analytical and experimental flexural behavior of concrete beams reinforced with glass fiber reinforced polymers bars. *Construction and Building Materials, 84*, 354–366.
2. Agarwal, B. D., Broutman, L. J., & Chandrashekhara, K., (2006). *Analysis and Performance of Fiber Composites* (3rd edn., Vol. 3, pp. 1–20) Wiley, Hoboken.
3. Alsayed, S. H., Al-Salloum, Y. A., Almusallam, T. H., & Amjad, M. A., (1999). *Concrete Columns Reinforced by Glass Fiber Reinforced Polymer Rods* (Vol. 188, pp. 103–112). Special Publication.
4. Bagheri, M., Chahkandi, A., & Jahangir, H., (2019). Seismic reliability analysis of RC frames rehabilitated by glass-fiber-reinforced polymers. *International Journal of Civil Engineering, 17*(11), 1785–1797.
5. Baker, A. A., Dutton, S., & Kelly, D., (2004). *Composite Materials for Aircraft Structures* (2nd edn., Vol. 1, pp. 1–23) AIAA, Reston.
6. Bhoopathi, R., Ramesh, M., & Deepa, C., (2014). Fabrication and property evaluation of banana-hemp-glass fiber-reinforced composites. *Procedia Engineering, 97*, 2032–2041.
7. Bond, I., Hucker, M., Weaver, P., Bleay, S., & Haq, S., (2002). Mechanical behavior of circular and triangular glass fibers and their composites. *Composites Science and Technology, 62*(7, 8), 1051–1061.
8. Bull, P. H., (2004). *Damage Tolerance and Residual Strength of Composite Sandwich Structures* (Vol. 1, pp. 1–20). PhD thesis, Royal Institute of Technology, Stockholm, Sweden.
9. Bunsell, A. R., & Renard, J., (2005). *Fundamentals of Fiber Reinforced Composite Materials* (Vol. 1, pp. 1–20). Institute of Physics Publishing Ltd., Bristol.
10. Bunsell, A. R., (1988). *Fiber Reinforcements for Composite Materials* (Vol. 1, pp. 1–23). Elsevier, Amsterdam.
11. Bunsell, A. R., (2005). Oxide fibers. In: Bansal, N. P., (ed.), *Handbook of Ceramic Composites* (Vol. 1, pp. 3–31). Kluwer Academic Publishers, Boston.
12. Case, J., Chilver, L., & Ross, C. T. F., (1999). *Strength of Materials and Structures* (4th edn., Vol. 4, pp. 1–24) Arnold, London.

13. Chan, Y., White, J. L., & Oyanagi, Y., (1978). A fundamental study of the rheological properties of glass-fiber-reinforced polyethylene and polystyrene melts. *Journal of Rheology, 22*(5), 507–524.
14. Chen, J., Zhao, D., Jin, X., Wang, C., Wang, D., & Ge, H., (2014). Modifying glass fibers with graphene oxide: Towards high-performance polymer composites. *Composites Science and Technology, 97*, 41–45.
15. Davies, P., Bigourdan, B., Chauchot, P., Choqueuse, D., Ferreira, A. J. M., Karjalainen, J. P., Hildebrand, M., et al., (2000). Predicting the behavior of a wide sandwich beam under pressure loading. *Proceedings of the 5th International Conference on Sandwich Construction* (Vol. 1, pp. 325–336). Zurich, Switzerland.
16. Deák, T., & Czigány, T., (2009). Chemical composition and mechanical properties of basalt and glass fibers: A comparison. *Textile Research Journal, 79*(7), 645–651.
17. Deitz, D. H., Harik, I. E., & Gesund, H., (2003). Physical properties of glass fiber reinforced polymer rebars in compression. *Journal of Composites for Construction, 7*(4), 363–366.
18. Deng, S., Ye, L., & Mai, Y. W., (1999a). Influence of fiber cross-sectional aspect ratio on mechanical properties of glass fiber/epoxy composites I. Tensile and flexure behavior. *Composites Science and Technology, 59*(9), 1331–1339.
19. Deng, S., Ye, L., & Mai, Y. W., (1999b). Influence of fiber cross-sectional aspect ratio on mechanical properties of glass fiber/epoxy composites II. Interlaminar fracture and impact behavior. *Composites Science and Technology, 59*(11), 1725–1734.
20. Dikbas, I., Gurbuz, O., Unalan, F., & Koksal, T., (2013). Impact strength of denture polymethyl methacrylate reinforced with different forms of e-glass fibers. *Acta Odontologica Scandinavica, 71*(3, 4), 727–732.
21. Ding, L., Liu, Y., Liu, J., & Wang, X., (2019). Correlation analysis of tensile strength and chemical composition of basalt fiber roving. *Polymer Composites, 40*(7), 2959–2966.
22. Drechsler, K., Middendorf, P., Van, D. B. B., & Heimbs, S., (2008). Advanced composite materials - technologies, performance, and modeling, In: Guedra, D. D., & Ladeveze, P., (ed.), *Course on Emerging Techniques for Damage Prediction and Failure Analysis of Laminated Composite Structures* (Vol. 1, pp. 147–197). Cepadues, Toulouse.

23. Eckerskorn, C., Mewes, W., Goretzki, H., & Lottspeich, F., (1988). A new siliconized-glass fiber as support for protein-chemical analysis of electroblotted proteins. *European Journal of Biochemistry, 176*(3), 509–519.
24. Fazilat, H., Ghatarband, M., Mazinani, S., Asadi, Z. A., Shiri, M. E., & Kalaee, M. R., (2012). Predicting the mechanical properties of glass fiber reinforced polymers via artificial neural network and adaptive neuro-fuzzy inference system. *Computational Materials Science, 58*, 31–37.
25. Georgiadis, S., Gunnion, A. J., Thomson, R. S., & Cartwright, B. K., (2008). Bird-strike simulation for certification of the Boeing 787 composite moveable trailing edge. *Composite Structures, 86*(1–3), 258–268.
26. Gessler, A., Maidl, F., & Schouten, M., (2007). Advancements in braiding technology for textile preforming. In: *28th SAMPE Europe International Conference* (Vol. 1, pp. 1–24). Paris.
27. Hao, Y., Liu, F., & Han, E. H., (2013). Protection of epoxy coatings containing polyaniline modified ultra-short glass fibers. *Progress in Organic Coatings, 76*(4), 571–580.
28. Havar, T., & Stuible, E., (2009). Design and testing of advanced composite load introduction structure for aircraft high lift devices. ICAF 2009, bridging the gap between theory and operational practice, *Proceedings of the 25th Symposium of the International Committee on Aeronautical Fatigue* (Vol. 25, pp. 365–374). Rotterdam, The Netherlands.
29. Hazen, J. R., & Dawson, D. K., (2007). *Aerospace Composites: A Design & Manufacturing Guide (Vol.1, pp. 10-34).*
30. Heimbs, S., Schmeer, S., Middendorf, P., & Maier, M., (2007). Strain rate effects in phenolic composites and phenolic impregnated honeycomb structures. *Composites Science and Technology, 67*(13), 2827–2837.
31. Heimbs, S., Strobl, F., Middendorf, P., & Guimard, J. M., (2010). Composite crash absorber for aircraft fuselage applications. In: *11th International Conference on Structures under Shock and Impact* (Vol. 11, pp. 1–24). Tallinn, Estonia.
32. Heimbs, S., Vogt, D., Hartnack, R., Schlattmann, J., & Maier, M., (2008). Numerical simulation of aircraft interior components under crash loads. *International Journal of Crashworthiness, 13*(5), 511–521.

33. Hucker, M., Bond, I., Bleay, S., & Haq, S., (2003). Experimental evaluation of unidirectional hollow glass fiber/epoxy composites under compressive loading. *Composites Part A, 34*(10), 927–932.
34. Humphrey, R. A., (1967). In: Shaped glass fibers. In: Broutman, L. J., & Krock, R. H., (eds.), *Modern Composite Materials* (Vol. 1, pp. 324–334), Addison-Wesley, Reading.
35. Jeong, J., Lee, Y. H., Park, K. T., & Hwang, Y. K., (2007). Field and laboratory performance of a rectangular shaped glass fiber reinforced polymer deck. *Composite Structures, 81*(4), 622–628.
36. Johnston, A. A., (1997). *An Integrated Model of the Development of Process Induced Deformation in Autoclave Processing of Composite Structures* (Vol. 1, pp. 1–24). PhD thesis, The University of British Colombia.
37. Jones, F. R., (2001). Glass fibers. In: Hearle, J. W. S., (ed.), *High-Performance Fibers* (Vol. 1, pp. 191–238). Woodhead Publishing Ltd., Cambridge.
38. Kalali, A., & Kabir, M. Z., (2012). Experimental response of double-wythe masonry panels strengthened with glass fiber reinforced polymers subjected to diagonal compression tests. *Engineering Structures, 39*, 24–37.
39. Karakaya, N., Papila, M., & Özkoç, G., (2020). Over-molded hybrid composites of polyamide-6 on continuous carbon and glass fiber/epoxy composites: 'An assessment of the interface'. *Composites Part A: Applied Science and Manufacturing, 131*, 105771.
40. Kouznetsova, V., Geers, M., & Brekelmans, W., (2004). Multi-scale second-order computational homogenization of multi-phase materials: A nested finite element solution strategy. *Computer Methods in Applied Mechanics and Engineering, 193*(48–51), 5525–5550.
41. Landesmann, A., Seruti, C. A., & Batista, E. D. M., (2015). Mechanical properties of glass fiber reinforced polymers members for structural applications. *Materials Research, 18*(6), 1372–1383.
42. Laura, D. M., Keskkula, H., Barlow, J. W., & Paul, D. R., (2002). Effect of glass fiber surface chemistry on the mechanical properties of glass fiber reinforced, rubber-toughened nylon 6. *Polymer, 43*(17), 4673–4687.
43. Li, H., Richards, C., & Watson, J., (2014). High-performance glass fiber development for composite applications. *International Journal of Applied Glass Science, 5*(1), 65–81.

44. Li, Y., Lin, Z., Jiang, A., & Chen, G., (2004). Experimental study of glass-fiber mat thermoplastic material impact properties and lightweight automobile body analysis. *Materials and Design, 25*(7), 579–585.

45. Lowrie, R. E., (1967). Glass fibers for high strength composites. In: Broutman, L. J., & Krock, R. H., (eds.), *Modern Composite Materials* (Vol. 1, pp. 270–323). Addison-Wesley, Reading.

46. Lu, H., Zhang, T., Wang, X. P., & Fang, Q. F., (2009). Electrospun submicron bioactive glass fibers for bone tissue scaffold. *Journal of Materials Science: Materials in Medicine, 20*(3), 793–798.

47. M€akinen, K., (1995). *Dynamically Loaded Sandwich Structures* (Vol. 1, pp. 1–33). Licentiate thesis, Royal Institute of Technology, Stockholm, Sweden, Report 95–2.

48. Mahrez, A., Karim, M. R., & Bt Katman, H. Y., (2005). Fatigue and deformation properties of glass fiber reinforced bituminous mixes. *Journal of the Eastern Asia Society for Transportation Studies, 6*, 997–1007.

49. Małek, M., Jackowski, M., Łasica, W., Kadela, M., & Wachowski, M., (2021). Mechanical and material properties of mortar reinforced with glass fiber: An Experimental Study. *Materials, 14*(3), 698.

50. Mallick, P. K., (2008). *Fiber-Reinforced Composites: Materials, Manufacturing and Design* (3rd edn., pp. 1–20). CRC Press, Boca Raton.

51. Mansor, M. R., Sapuan, S. M., Zainudin, E. S., Nuraini, A. A., & Hambali, A., (2013). Hybrid natural and glass fibers reinforced polymer composites material selection using analytical hierarchy process for automotive brake lever design. *Materials & Design, 51*, 484–492.

52. Middleton, D. H., (1990). *Composite Materials in Aircraft Structures* (Vol. 1, pp. 1–32). Longman, New York.

53. Moustafa, A., & ElGawady, M. A., (2016). Strain rate effect on properties of rubberized concrete confined with glass fiber-reinforced polymers. *Journal of Composites for Construction, 20*(5), 04016014.

54. Nam, Y. W., Choi, J. H., Lee, W. J., & Kim, C. G., (2017). Fabrication of a thin and lightweight microwave absorber containing Ni-coated glass fibers by electroless plating. *Composites Science and Technology, 145*, 165–172.

55. Palanikumar, K., Karunamoorthy, L., & Karthikeyan, R., (2006). Assessment of factors influencing surface roughness on the machining of glass fiber-reinforced polymer composites. *Materials & Design, 27*(10), 862–871.

56. Park, S. J., & Jin, J. S., (2001). Effect of silane coupling agent on interphase and performance of glass fibers/unsaturated polyester composites. *Journal of Colloid and Interface Science, 242*(1), 174–179.

57. Pierard, O., Friebel, C., & Doghri, I., (2004). Mean-field homogenization of multi-phase thermo-elastic composites: A general framework and its validation. *Composites Science and Technology, 64*(10, 11), 1587–1603.

58. Praeger, M., Saleh, E., Vaughan, A., Stewart, W. J., & Loh, W. H., (2012). Fabrication of nanoscale glass fibers by electrospinning. *Applied Physics Letters, 100*(6), 063114.

59. Puska, M. A., Lassila, L. V., Närhi, T. O., Yli-Urpo, A. U., & Vallittu, P. K., (2004). Improvement of mechanical properties of oligomer-modified acrylic bone cement with glass-fibers. *Applied Composite Materials, 11*(1), 17–31.

60. Rotem, A., & Lifshitz, J. M., (1971). Longitudinal strength of unidirectional fibrous composite under high rate of loading. *Proceedings of the 26th Annual Technical Conference, Society of the Plastics Industry, Reinforced Plastics/ Composites Division* (pp. 1–10). Washington, D. C., Section 10-G.

61. Samal, S. K., Mohanty, S., & Nayak, S. K., (2009). Polypropylene—bamboo/glass fiber hybrid composites: Fabrication and analysis of mechanical, morphological, thermal, and dynamic mechanical behavior. *Journal of Reinforced Plastics and Composites, 28*(22), 2729–2747.

62. Sanghera, J. S., Shaw, L. B., & Aggarwal, I. D., (2009). Chalcogenide glass-fiber-based mid-IR sources and applications. *IEEE Journal of Selected Topics in Quantum Electronics, 15*(1), 114–119.

63. Sathishkumar, T. P., Satheeshkumar, S., & Naveen, J., (2014). Glass fiber-reinforced polymer composites: A review. *Journal of Reinforced Plastics and Composites, 33*(13), 1258–1275.

64. Shuaib, N. A., Mativenga, P. T., Kazie, J., & Job, S., (2015). Resource efficiency and composite waste in UK supply chain. *Procedia CIRP, 29*, 662–667.

65. Sierakowski, R. L., (1997). Strain rate effects in composites. *Applied Mechanics Reviews, 50*(12), 741–761.

66. Sims, G. D., & Broughton, W. R., (2000). Glass fiber reinforced plastics-properties. In: *Comprehensive Composite Materials, Polymer Matrix Composites* (Vol. 2, pp. 151–197). Elsevier, Amsterdam.

67. Soares, P. A., Batalha, M., Souza, S. M. G. U., Boaventura, R. A., & Vilar, V. J., (2015). Enhancement of a solar photo-Fenton reaction with ferric-organic ligands for the treatment of acrylic-textile dyeing wastewater. *Journal of Environmental Management, 152*, 120–131.

68. Van, T. M. J. L., (1998). *Sandwich Fuselage Design* (Vol. 1, pp. 12–33) Delft University Press, Delft.

69. Welsh, L. M., & Harding, J., (1998). Effect of strain rate on the tensile failure of woven reinforced polyester resin composites. *Journal de Physique, 46*(8), C5.405–C5.414.

70. Witten, E., (2010). *The European Composites Market 2008/2009* (pp. 16–18). JEC Composites Magazine, No. 55.

71. Wu, H. F., Dwight, D. W., & Huff, N. T., (1997). Effects of silane coupling agents on the interphase and performance of glass-fiber-reinforced polymer composites. *Composites Science and Technology, 57*(8), 975–983.

72. Wu, S. H., Wang, F. Y., Ma, C. C. M., Chang, W. C., Kuo, C. T., Kuan, H. C., & Chen, W. J., (2001). Mechanical, thermal, and morphological properties of glass fiber and carbon fiber reinforced polyamide-6 and polyamide-6/clay nanocomposites. *Materials Letters, 49*(6), 327–333.

73. Xia, Y., & Wang, X., (1996). Constitutive equation for unidirectional composites under tensile impact. *Composites Science and Technology, 56*(2), 155–160.

74. Xia, Y., Yuan, J., & Yang, B., (1994). A statistical model and experimental study of the strain-rate dependence of the strength of fibers. *Composites Science and Technology, 52*(4), 499–504.

75. Xu, M., Yang, X., Zhao, R., & Liu, X., (2013). Copolymerizing behavior and processability of benzoxazine/epoxy systems and their applications for glass fiber composite laminates. *Journal of Applied Polymer Science, 128*(2), 1176–1184.

76. Yao, Y., Zeng, X., Guo, K., Sun, R., & Xu, J. B., (2015). The effect of interfacial state on the thermal conductivity of functionalized Al_2O_3 filled glass fibers reinforced polymer composites. *Composites Part A: Applied Science and Manufacturing, 69*, 49–55.

Chapter

5 Properties, Fabrication, and Applications of Carbon Fiber-Reinforced Polymers

CONTENTS

5.1. Introduction 160

5.2. Manufacturing Processes 161

5.3. Surface Treatment 162

5.4. Materials for CFRP 165

5.5. Surface Treatment of the Carbon Fibers 166

5.6. Preparation of Carbon Fiber-Strengthened Polymer Composites 170

5.7. Characterization of the Carbon Fibers 170

5.8. Characterization of CFRPCS 171

5.9. Electrochemical Oxidation Effect 176

5.10. Electroplating Effect 179

5.11. Applications 186

References 189

5.1. INTRODUCTION

Nowadays, carbon fibers are amongst the most essential reinforcing materials characterized by very high modulus and strength as well as high-temperature resistance. The scientific efforts have been directed towards analyzing and improving their performance in certain composite systems (Paiva et al., 2000; Morgan, 2005). In spite of these efforts and advantages, it is known very well that the interfacial adhesion amongst matrix and carbon fibers in the composite system is very poor to guarantee better mechanical performance (Wu, 1982; Schwartz, 1983).

In the composite system, the adhesion degree at an interface between matrix and fiber plays a vital role in enhancing the consequential mechanical behavior because of the presence of secondary or physical interaction, that is, hydrogen force and the van der Waals attraction at an interface. Moreover, the adhesion degrees at interfaces are important for the durability and strength of the composites made from these materials (Park et al., 2002; Yosomiya, 2020).

Polymeric composites like epoxy resins/ carbon fibers and PEEK systems/ carbon fibers are now being utilized in various aerospace, recreational, and marine applications. The specific composites selected depend on the use. Epoxy resins have been demonstrated to be very versatile as the resin can be cross-linked with many different anhydrides, acids, and amines. They can react with numerous other polymer substances (Clayton, 1988; Ho et al., 2007).

However, the properties of the composites are not only governed by properties of the individual constituents but also by an interface between the matrix resins and the fibers. Successful reinforcement of the composite materials is accomplished by only attaining adequate stress transfer between matrix resins and the fibers. This can be comprehended by chemical and physical adhesion between the matrix resins and the fibers. Epoxy resins do not bond strongly with the untreated carbon fibers. In order to solve this issue, pretreatments, generally having oxidative nature, of the fiber surfaces have normally been developed that enhance the matrix/fiber adhesion (Meservey and Tedrow, 1978).

This chapter emphasizes the surface modification of PAN (polyacrylonitrile)-based carbon fibers utilizing different methods like plasma and electrochemical techniques to control the upsurge in surface roughness to improve the matrix/fiber interface in the CFRCs (carbon-fiber-reinforced polymer composites) (Table 5.1).

Table 5.1. Distinctive Properties of Three Grades of the Carbon Fibers

Typical Properties	High Strength	High Modulus	Intermediate Modulus	Ultrahigh Modulus
Young's modulus (GPa)	230	370	290	520–620
Density (g/cm^3)	1.8	1.9	1.8	2.0–2.1
Tensile elongation (%)	1.5	0.7	1.9	0.8
Tensile strength (GPa)	3.5	2.2	5.5	3.4

5.2. MANUFACTURING PROCESSES

There are numerous manufacturing methods for producing the structural products of composite, with several variations and patented procedures like lay-up, winding, molding, and pultrusion. Among them, four basic manufacturing techniques are addressed (Rapley et al., 1990):

- The lay-up procedure engages a machine or hand buildup fiber mats that are generally held together perpetually by the resin system. This approach enables several layers of diverse fiber alignments to be made up to an anticipated product shape and sheet thickness.
- Molding is the strip of material having several cross-sections utilized to cover transitions amongst surfaces or for decoration. Molding is amongst the most common approaches to shaping plastic resins.
- The process of filament winding can be automated in order to swathe resin-wetted fibers around the mandrel to yield polygonal or circular shapes.
- The process of pultrusion involves the constant pulling of mats and fiber rovings through the resin bath and afterward into the heated die. The high temperature within the die preserves the composite matrix into the persistent cross-sectional structural shape (Curran et al., 2016; Ling et al., 2019).

5.3. SURFACE TREATMENT

In the early 1970s and late 1960s, various kinds of surface treatment were utilized (Jones, 1991, 1993; Delamar et al., 1997):

1. **Wet Methods:** Fibers' immersion in the oxidizing agents like chromic and nitric acid and electrochemical oxidation. The electrochemical oxidation method is the widely used industrial approach for treating carbon fibers (Nakao et al., 1992).
2. **Dry Methods:** Oxidation in oxygen or air. Plasma treatment is amongst the effective and convenient methods to yield functional groups on the surfaces of carbon fiber.

The chemical and physical properties of the carbon fibers are normally modified to accomplish better adhesion between the matrix and the reinforcement materials. To increase the surface polarity of the nonpolar carbon fibers, several surface treatment methods are applied, like oxidation in acid solutions, anodic oxidation, dry oxidation in oxygen, mild fluorination, plasma treatments, and metallic coating. The interfacial adhesion between matrices and the fibers cannot be accomplished without close contact, that is, till the surface of fiber contacts a resin on the level of intermolecular equilibrium distance (Wang and Garton, 1992; Wu et al., 1995).

The large interfacial area of close contact by a matrix resin is the necessity every time the interfacial bond is developed mainly due to the van der Waals adsorption force. Additionally, complete fibers' wetting with the help of a liquid polymeric matrix is beneficial when the free energy of the surface of the fiber is well above as compared to that of a liquid polymer in observation of London dispersive-particular polar components of the surface free energy (Fukunaga and Ueda, 2000; Montes-Morán et al., 2005).

Untreated carbon fibers available for commercial use have a surface tension near to 40 mJm^{-2} and a tremendously hydrophobic nature because of the very high temperature of the manufacturing process triggering graphitization (2500°C) or carbonization (700°C to 1500°C). Conversely, most of the polymeric matrix resins, like phenolic (PF) resin, are somewhat hydrophilic with a surface tension of 35 to 45 mJm^{-2} (Miyahira and Soule, 2019).

It is well known that the surface treatments enhance the bonding of fiber-resin, but the procedure by which the interfacial adhesion is enhanced is still debatable. There seem to be three main mechanisms that contribute to increasing the strength of fiber-resin bond (Chong and Ohara, 1992; Oliver et al., 2000):

- The adsorption (chemical or physical) of the molecules of resin onto the surface complexes; generally, acidic complexes have been well-thought-out to be of utmost importance.
- Elimination of pollutants providing a better surface for adhesion.
- The effect of mechanical interlocking is activated by the oxidation groups when the resin enters channels and pits on the coarsened surface, the fiber surface also interacts chemically with the epoxy resin, resulting in bonding between resin and fiber (Table 5.2) (Park and Kim, 2000).

Table 5.2. Description of Some Carbon Fibers

Product Name	Manufacturer	Precursor Type	Properties
HMG-50	Hitco	Rayon	High modulus
Thornel-25	Union carbide	Rayon	High strength
AC	Le Carbone Lerraine	PAN	High strength
AG	Le Carbone Lerraine	PAN	High modulus
Type II	Not reported	PAN	High strength
Type I	Not reported	PAN	High modulus
HMS	Hercules Inc.	PAN	High modulus
HMU	Hercules Inc.	PAN	High modulus
P100	Amoco	Pitch	High modulus
T300	Amoco	PAN	High strength

In most cases, an upsurge in the functionality of the surface is escorted by an upsurge in the surface area, and therefore the comparative significance of these two factors is quite difficult to evaluate. Pittman et al. (1998) reported that treatment of the Thornel-25 fibers in the nitric acid upsurges the surface area and the amount of surface oxygen (O_2) complexes. The IFSS (interfacial shear strength) of the composites made from treated fibers was also augmented. Figure 5.1 displays the functional groups that are considered to be existent on the oxidized surfaces of carbon. One or more functional groups can also be present of the surface of the fiber.

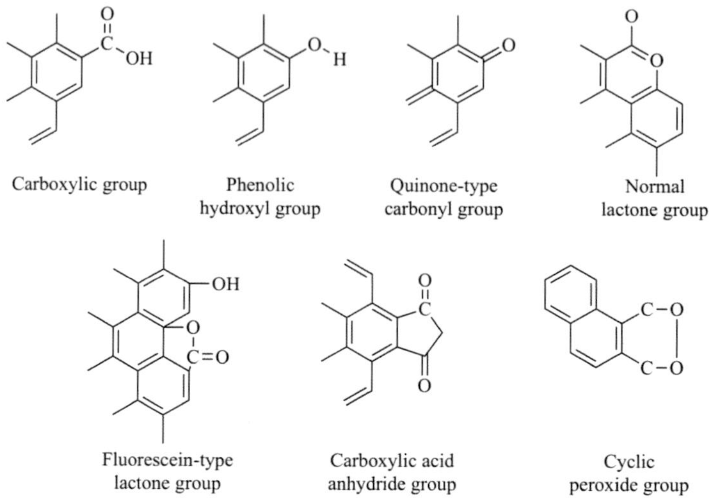

Figure 5.1. The probable functional groups introduced on the oxidized surfaces of carbon.

Source: https://courses.lumenlearning.com/boundless-chemistry/chapter/functional-group-names-properties-and-reactions/.

Some of the fibers, after treatment with nitric acid, were heated afterward in hydrogen to decrease the surface oxygen(O_2) complexes. The IFSS of the conforming composites decreased considerably, signifying that the functionality of the surface was of utmost significance in improving fiber-resin bonding. Park et al. (2002) found no reduction in IFSS for composites formed from the treated Hitco HMG-50 fibers (Table 5.1) that had usually been heated in hydrogen before assimilation into the composite.

It has been discovered by numerous researchers that by heating the fiber acidic surface oxidation is removed almost entirely. The AG treatment (Table 5.1) carbon fibers in the Hummers reagent yields the graphitic oxide layer (Hage et al., 1997; Beydon et al., 2000). The lamellar graphitic oxide layer can be damaged by heating. The creation of the lamellar coat improves the IFSS of the composites, but the IFSS made from the treated fibers decreased to an analogous value on heating as compared to that of untreated fiber-resin composites. For AC (Table 5.1) fibers, heating did not produce such a noticeable effect. Even though heating instigated a reduction in the IFSS, the original value for untreated fiber-resin composites was not regained. Delamar et al. (1997) examined the effect of treating fibers electrochemically in the nitric acid on the shear strength of the composites.

They utilized titration approaches to conclude the concentration of surface oxide and the blocking reagents like diazomethane. Over oxidation of the fibers was discovered to impact composite fracture behavior, instigating fragile fracture in a short-beam shear test. Keeping this thing in mind, only short times of oxidation times used. They determined that the bonding amount of fiber-resin only depends upon the quantity of acidic functionality of the surface. Surprisingly, Weis and Fitzer discovered that the composites developed from type II and I fibers (5.1) provided the same values of IFSS, but type I had the concentration of surface oxide 1 order of magnitude less as compared to type II fibers. This certainly specifies that surface treatments encourage bonding through some other mechanism along with the one they recommended (Hoecker and Karger-Kocsis, 1993; Li et al., 2009).

Brelant (1981) has an entirely different view on this subject, signifying the distribution and size of voids at an interface of fiber-resin. This limiting strength is the stress developed to spread cracks through the holes along the boundary of the fiber matrix. Brelant suggested that the upsurge in shear strength for short treatment times might be accredited to the elimination of pollutants or reduction of the protuberances that help void creation. He recommended that the bonding of fiber-resin is mainly physical in nature.

This was confirmed later by Drzal and his coworkers (1983) who published proof to exhibit that the effect of the commercial treatment was to eliminate feebly bound crystallites from an untreated fiber surface and therefore offer the better surface to which this resin could adhere. Kozlowski and Sherwood (1985, 1986) discovered that the crystallites were eliminated by diverse mechanisms dependent on the pH of the electrolyte used. In the basic solutions, small pieces of fiber were discovered in the operational electrode solution; while in the acidic solutions, they were generally oxidized to CO_2 (carbon dioxide).

5.4. MATERIALS FOR CFRP

The carbon fibers utilized in this particular study were unsized and untreated PAN-centered fibers with high strength, TZ-307(12 K monofilaments), developed by the Taekwang Company of Korea. The average diameter of the carbon fibers was roughly 7 mm, and the usual strength and tensile modulus were about 3.5 and 245 GPa, correspondingly (Park, 1999; Park et al., 2000).

The epoxy resin utilized in this particular study was DGEBA (diglycidylether of bisphenol-A) YD-128 supplied by the Kukdo Chemical

Company located in Korea. The epoxide equal weight was 185 to 190 g equiv^{-1}, and the viscosity and density were 5000 cPs and 1.16 g cm^3, correspondingly, at 25°C. DDM (Diaminodiphenylmethane) supplied by the Aldrich Chemical Company were chosen as the hardener for curing, and MEK (methyl ethyl ketone) was used to decrease the DGEBA's high viscosity. Resol-kind PF resin (CB-8057, made by Kangnam Chemical Company located in Korea) was also utilized as a polymeric matrix (Kennedy and Frese, 1978; Park and Lee, 1998).

5.5. SURFACE TREATMENT OF THE CARBON FIBERS

5.5.1. Electrochemical Oxidation

Electrochemical treatment of the carbon fibers was carried out utilizing the laboratory apparatus that was described earlier to present the functional groups and also to enhance the adhesion degrees at interfaces between matrix resins and fibers. The electrolyte utilized was 10 wt.% phosphoric acid solutions with a constant oxidation rate. The utilized electric current densities were 1.6, 0.8, 0.4, 0.2 and 0 A/m^2. After the anodic oxidation, the carbon fibers anodized were washed with fresh distilled water and cleaned with acetone in the Soxhlet extractor for two hours to eliminate residual oxides or surface impurities. The laboratory plant is schematically demonstrated in Figure 5.2 (Morra et al., 1991; Pittman et al., 1998).

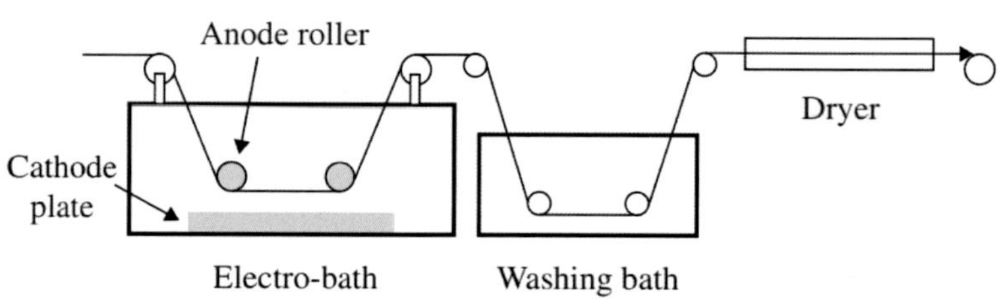

Figure 5.2. Schematic representation of constant electrolytic surface anodization procedure.

Source: https://onlinelibrary.wiley.com/doi/book/10.1002/9783527645213.

5.5.2. Electroplating

Electroplating has been utilized to yield metal matrix composites strengthened with carbon fibers. The surfaces of carbon fiber are metalized via electrolysis in molten salt solutions. The electroplating device was made that can uninterruptedly plate nickel (Ni) onto the surfaces of fiber. The carbon fibers speed was governed by the gearbox and the speed was usually about 0.72 m min^{-1}. Nickel sulfate was the key salt utilized in the electroplating solution, and the electrolytic Ni plate was utilized as an anode. Carbon fibers are generally conductive; hence they were directly utilized as a cathode in order to be plated. The operating conditions and compositions of the plating bath are provided in Table 5.3. Before plating, these carbon fibers were stimulated in nitric acid for 30 minutes in order to improve the interfacial adhesion between carbon fibers and the nickel coating. Before utilizing fiber surface examination or the preparation of composites, the remaining chemicals were also eliminated by Soxhlet extraction with acetone at 70°C for two hours. Lastly, the carbon fibers were cleaned numerous times with the distilled water and then dried in the vacuum oven at 60°C for 12 hours (Proctor and Sherwood, 1982; Wang, 1995).

Table 5.3. Operating Conditions and Composition of a Nickel-Plating Bath

Compositions	$NiSO_46H_2O$	280 g/L
	H_3BO_3 $NiCl_26H_2O$	30 g/L 40 g/L
Conditions	pH	4.5–5.0
	Current density (A/m^2) Temperature (°C)	0–60 45 1

5.5.3. Oxyfluorination

For the carbon fibers, the oxygen-comprising fluorination treatment gives several benefits over the other treatments. One main benefit is that the fibers' mechanical properties are not considerably degraded if optimal conditions are operated. Some other important benefits are associated with concerns regarding the advancement of environment-friendly procedures. These carbon fibers were imperiled to oxyfluorination under diverse conditions (Kinloch, 1987; Neumann et al., 2010). The reaction of oxyfluorination was performed in the batch reactor composed of nickel with the external electric furnace, as displayed in Figure 5.3. After evacuation, the oxygen and fluorine mixtures were presented to a reactor at room temperature, and the reactor was heated afterward to the temperature of treatment. After this reaction, the samples were cooled down to room temperature, and the reactive gases were then eliminated from the reactor with nitrogen gas. In the circumstance of the room temperature reaction, this reactor was cooled down and evacuated in the cooling bath before the charging of fluorine.

Table 5.4. Experimental Oxyfluorinated Conditions of the PAN-Centered Carbon Fibers Utilized

Specimens	F_2/O_2 Mixtures (%)	Oxyfluorination Temperature (°C)
No treatment	–	–
CFO-RT	50/50	25
CF-RT	100/0	25
CFO-300	–	300
CFO-100	–	100
CFO-400	–	400

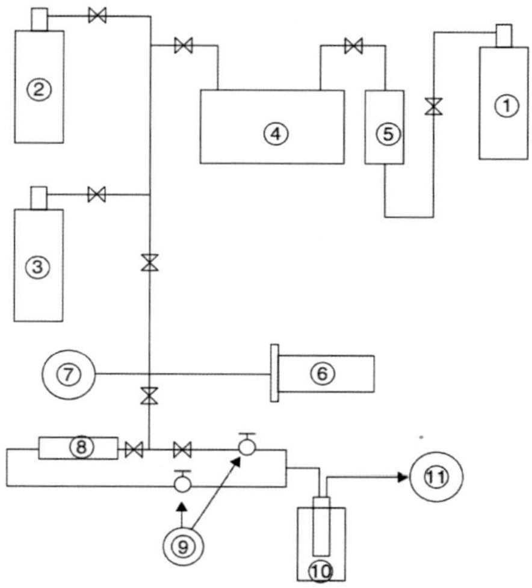

Figure 5.3. Schematic representation of the oxyfluorination reactor: (1) F_2 (fluorine) gas cylinder; (2) N_2 (nitrogen) gas cylinder; (3) O_2 (oxygen) gas cylinder; (4) the buffer tank; (5) HF absorber; (6) reactor; (7) pressure gauge; (8) fluorine absorber (aluminum oxide); (9) glass cock; (10) nitrogen in liquid form; and (11) the rotary vacuum pump.

Source: https://www.researchgate.net/publication/285993757_Carbon_Fiber-Reinforced_Polymer_Composites_Preparation_Properties_and_Applications.

The reactor was eliminated from the bath of cooling after fluorine was eliminated with nitrogen. The overall gas pressure was nearly 0.2 MPa and the minimal reaction time was 10 minutes at the temperature of treatment. Table 5.4 gives the conditions of an experiment of the studied carbon fibers (Severini et al., 2002; Mallick, 2007).

5.5.4. Plasma Modification

Plasma treatment is amongst the most applied surface modifications for the commercial manufacturing of carbon fibers. These carbon fibers were generally exposed to the atmosphere argon plasma at the mercy of the subsequent conditions: 150 W power, 13.56 MHz radiofrequency, 10 millibar pressure at almost 50 sccm min^{-1} utilizing the devices of atmospheric pressure plasma equipped with MFC (mass flow controller) hardware, as exhibited in Figure 5.4 (Owens and Wendt, 1969; Kumar et al., 2007).

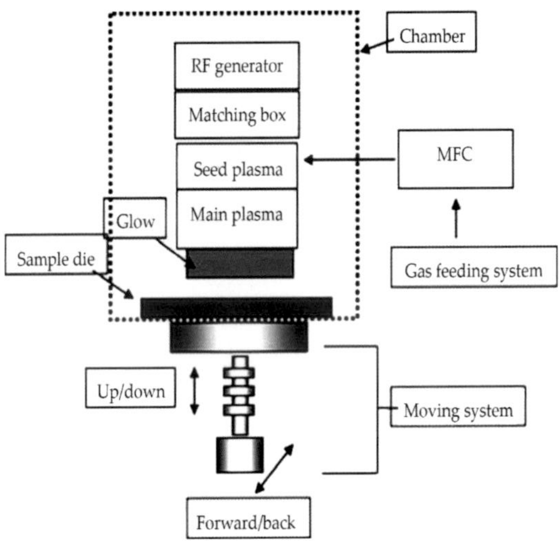

Figure 5.4. Schematic representation of the atmospheric pressure plasma devices.

Source: https://www.semanticscholar.org/paper/Conference-ISSS-7-Synthesis-of-SiOC-(-%E2%80%93-H-)-Films-Mori-Masuko/236f1383faa644c904d833a3ac13a07e8426a1a7/figure/0.

The exposure time was changed from 0–180 minutes and they were designated as:

- CFs-0;
- CFs-10;
- CFs-20;

- CFs-30; and
- CFs-180, correspondingly.

5.6. PREPARATION OF CARBON FIBER-STRENGTHENED POLYMER COMPOSITES

UD composite laminates were made by constant impregnation of fibers utilizing the technique of drum winding for making prepregs with following hot pressing. The laminates prepared with 32 plies of the prepregs were made up in the hot press at nearly 7.4 MPa at 150°C for 150 minutes with the vacuum bagging approach in the conventional composite processing (Jensen et al., 2003; Kawai et al., 2001). In order to attain the average V_f (fiber volume fraction) of composites, the small-size rectangular sample was sliced from the laminate, and then the side section of the sample was polished. The N number of fibers was totaled utilizing video images of the small area 0.2 mm (width direction), 0.1 mm (thickness direction) and d_f (average diameter) of fiber was measured at the cross-section. The V_f was calculated from Eqn. (1):

$$V_f = \frac{N \cdot d_f^2 \cdot \pi}{4S} \quad (1)$$

The V_f of bulk samples was about 52% (0.2%) for all of the composites.

5.7. CHARACTERIZATION OF THE CARBON FIBERS

To study the surface treatment effect of the carbon fibers, it is essentially helpful to conclude the surface treatments' effect on the chemistry of fiber surfaces. The XPS (X-ray photoelectron spectroscopy) measurement of the surfaces of fiber was performed using the VG Scientific ESCA LAB MK-II spectrometer fortified with the Magnesium-Ka X-ray source. In the specimen chamber, the base pressure was governed in between 10^8 and 10^9 torr. In a survey spectrum (0 eV to 1100 eV), the XPS program package made the C1s peak to fix at 284.6 eV bearing in mind the neutralization. For high-resolution spectra of C1s, the accurate position of the C1s peak was not fixed (Park and Kim, 2005). Quantification was conducted by applying the standard procedure, that is, the linear background subtraction, the measured lines to be fitted by means of the Gaussian curves set, and transforming the intensities into the atomic concentrations by utilizing comparative sensitivity factors (Park et al., 2003a, b).

Conversely, carbon fibers are problematic to work with as the carbon fibers do not lend themselves willingly to the common spectroscopic methods owing to their size, color, and handling complications. Even with XPS, data analysis has been demonstrated to be problematic owing to the very small concentration of the chemical species existent on the surfaces of fiber. Thus, the contact angle method was performed to inspect the surface conditions of fibers (Ryu et al., 1999; Lee et al., 2000).

5.8. CHARACTERIZATION OF CFRPCS

5.8.1. ILSS (Interlaminar Shear Strength)

Interlaminar shear strength (ILSS) was measured in order to approximate the composites' interfacial adhesion strength. The test was carried out by 3 point short-beam bending tests approach utilizing the Imodel Lloyd LR-5K mechanical tester conferring to ASTM D 2344. For the rectangular cross-section of composites, the ILSS concluded from the 3-point bending test is calculated with Eqn. (2) (Figure 5.5).

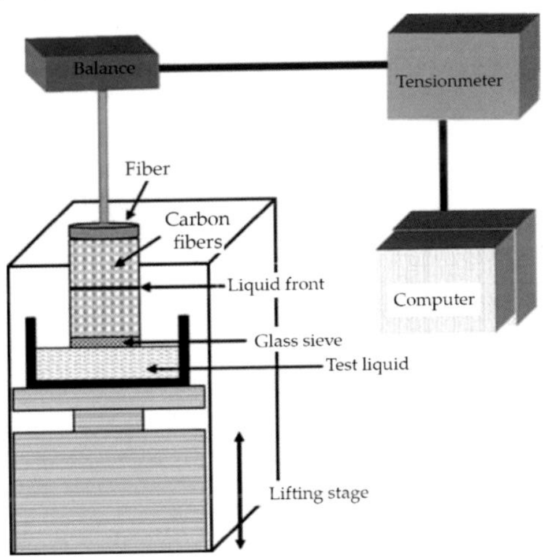

Figure 5.5. Schematic representation of the principles of capillary rise technique.

Source: https://onlinelibrary.wiley.com/doi/abs/10.1002/9783527645213.ch5.

$$\text{ILSS} = \frac{3P}{4bd} \qquad (2)$$

where; P is the load at the moment of breakage; *b* is the width of the sample; *d* is the thickness of the sample.

The range and accuracy of a load cell utilized were 5 kN and 0.5% grade, correspondingly. The composites were generally machined along the direction of fiber into 6 x 30 mm short-beam shear samples with five mm thickness (Figure 5.6 and Table 5.5).

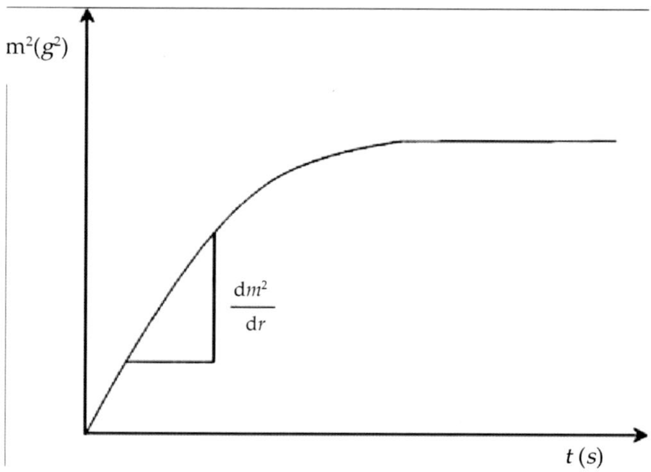

Figure 5.6. *Usual data of wicking rate in weight m^2 as the function of time (t).*

Source: https://onlinelibrary.wiley.com/doi/10.1002/9783527645213.ch5.

Table 5.5. Features of the Wetting Liquids Utilized in This Work

Wetting liquids	**c^L_L (mJ/m^2)**	**c_L (mJ/m^2)**	**cSPL [mJ/m^2]**	**g (MPa s)**	**r (g/cm^3)**
Water	21.8	72.8	51	1	0.998
n-Hexane	18.4	18.4	0	0.33	0.661
Ethylene glycol	31.0	47.7	16.7	17.3	1.100
Diiodo-methane	50.42	50.8	0.38	2.76	3.325

Note: c^{SP}_L: specific component of the free energy of the surface; c^L_L: London dispersive component of the free energy of the surface; c_L: total free energy of the surface; r: density; g: viscosity.

The distance amongst supports divided by the thickness of the sample's L/d=5, and the speed of the crosshead was fixed at 2 mm min^{-1} (Paiva et al., 2000; Lawton et al., 2005).

5.8.2. Critical Stress Intensity Factor

The K_C (critical stress intensity factor), which is amongst the parameters of fracture toughness, was defined by a state of the stress in the locality of the crack tip as the function of sample geometry, the geometry of crack, and the load applied on basis of the linear elastic fracture mechanics (LEFM). An analytical expression for K_C in the mode I fracture (K_{IC}) was characterized utilizing the SEN (single-edge notched) beam fracture test of toughness in a 3 point-bending flexure, as displayed schematically in Figure 5.7. Notches were sliced using the diamond wire saw with slow speed, roughly half the depth of a specimen. The SEN beam fracture test of toughness was carried out using the Instron Model 1125 tester conferring to ASTME 399 (Schwartz, 1983; Nakajima, 2000). For the SEN beam fracture test of toughness, the K_{IC} value is calculated with the help of Eqn. (3).

$$K_{IC} = \frac{3PLa^{1/2}}{2b^2d}Y \qquad (3)$$

where; P is the critical load for the propagation of crack (N); L is the length of a span(mm); a is the pre-crack length (mm); b is the width of the specimen (mm); d is the thickness of specimen (mm); Y is the geometrical factor calculated by Eqn. (4):

$$Y = 1.98 - 3.07(a/b) + 14.53(a/b)^2 - 25.11(a/b)^3 + 25.80(a/b)^4 \qquad (4)$$

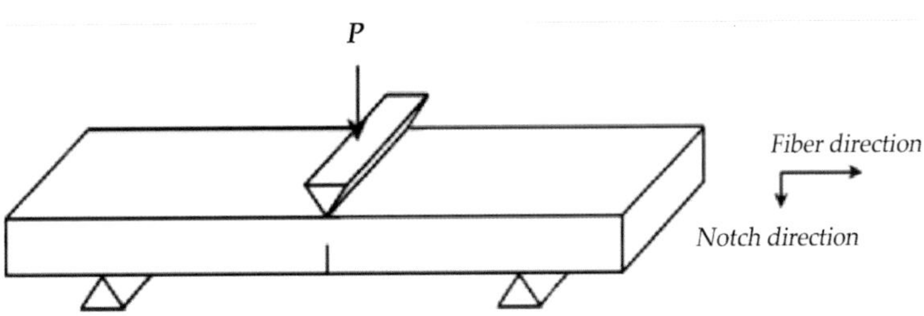

Figure 5.7. Schematic representation of SEN (single-edge notched) beam fracture test of toughness.

Source: https://www.researchgate.net/figure/Single-edged-notched-beam-test_fig2_227846644.

5.8.3. Mode I Interlaminar Factor of Fracture Toughness

Mode I G_{IC} (interlaminar fracture toughness) was concluded according to the ASTM D 5528-94 utilizing the DCB (double cantilever beam) test, as displayed in Figure 5.8. Samples were sliced to size using the diamond saw with the direction of fiber parallel to the sample length, 3 millimeters in thickness, 20 millimeters in width, and 160 millimeters in length with the 30-millimeter film of crack starter. Aluminum tabs were fixed to the ends of samples comprising the insert using the high strength epoxy adhesive and permitted to dry for almost 48 hours. The Instron 1125 Universal Testing Machine was utilized to carry out the tests using the crosshead speed of one mm min^{-1} and the load vs displacement curve was made for each of the tests (Chwastiak, 1973; Michalski and Saramago, 2000). From the data attained for these tests, the mode I rate of critical strain energy release was calculated utilizing modified beam theory conferring to Eqn. (5):

$$G_{IC} = \frac{3P \cdot \delta}{2W(a + |\Delta|)} \tag{5}$$

where; P is the load; d is the displacement; W is the specimen width; a is the crack length; |D| is the crack length correction factor.

G_{IC} values were plotted as the function of a (crack length) to produce the resistance R curve.

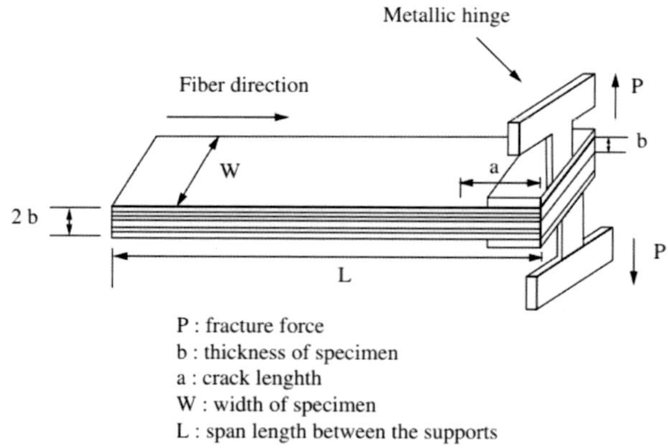

Figure 5.8. Schematic representation of DCB (double cantilever beam) specimen.

Source: https://www.researchgate.net/figure/Mode-I-Double-Cantilever-Beam-DCB-specimen-reprinted-from-8_fig4_328464812.

5.8.4. Fracture Behaviors

Ahearn and Rand (1996) exhibit the 3 general classes of the fracture behavior outcome, dependent on the comparative magnitudes of interfacial bond fracture energy: the elastic disparity between matrix and fiber, the structural energy of fiber, and sliding resistance between de-bonded matrix and fiber. The common shapes of load-deflection curves produced from these 3 kinds of behavior are displayed in Figure 5.9 (Park and Kim, 2001; Seo et al., 2005).

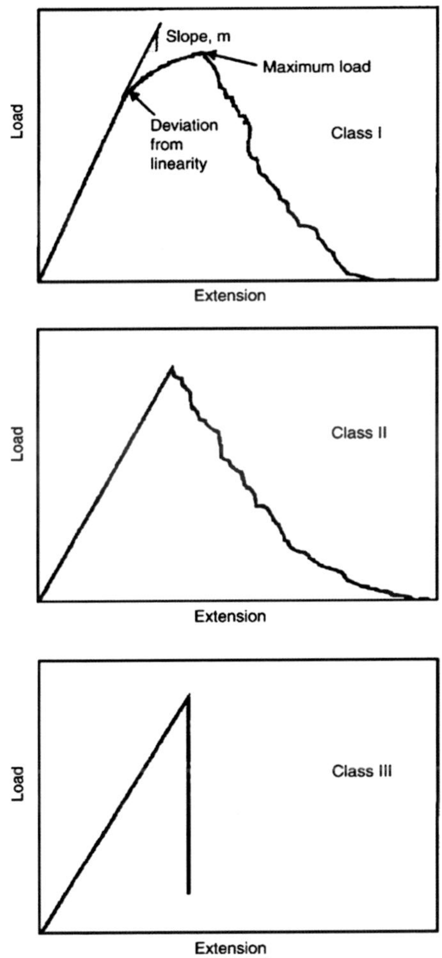

Figure 5.9. Load-extension curves produced by three key classes of the fracture behaviors.

Source: https://www.researchgate.net/publication/285993757_Carbon_Fiber-Reinforced_Polymer_Composites_Preparation_Properties_and_Applications.

It is observed that de-bonding happens triggering class I-type behavior, particularly in the case of malleable matrix composites. Now, class I is the conventional multiple fracture mode of the failure, leading to critical strength being controlled largely by the properties of fiber, and ascends when interfacial sliding stress is quite low in order to permit comparative motion of the fiber (Table 5.6).

Table 5.6. Variation of the Surface Chemical Compositions of Carbon Fibers with the Current Density Perceived by XPS Measurement (at Percent)

Current Density (A/m^2)	XPS			
	C_{1s}	O_{1s}	N_{1s}	O_{1s}/C_{1s}
1.6	78.0	14.2	7.8	18.2
0.8	76.9	14.0	9.1	18.2
0.4	79.2	13.1	7.7	16.5
0.2	80.7	12.2	7.1	15.1
0	91.8	6.9	1.3	7.5

On the other hand, class III-type behavior takes place with the usual brittle matrix composites. De-bonding occurs and the composite flops terribly (Ahearn and Rand, 1996; Gopal et al., 1997).

5.9. ELECTROCHEMICAL OXIDATION EFFECT

5.9.1. Surface Characteristics

It is known very well that XPS is a very beneficial apparatus in the establishment of functional groups and chemical composition of the surfaces of fiber. As exhibited in Table 5.6, the amounts of surface nitrogen and oxygen groups augmented with an upsurge in current density in the electrolytic solution. The large fluctuation of the oxygen group is because of the changing acidity of surface functional groups. The C_{1s}/O_{1s} ratios as the function of the electric potential applied are also given in Table 5.6, and the ratios upsurge with the growing electric current density equal to about 0.4 A m^{-2}. When the ratio of O_{1s}/C_{1s} is well-thought-out, it is very clear that during the electrochemical oxidation, the carbonyl (C=O) and the hydroxyl (COOH) groups augmented on the surfaces of fiber, as observed in Table 5.7 (Proctor and Sherwood, 1982; Denison et al., 1986).

Table 5.7. Fluctuation of the Functional Groups of the Surfaces of Carbon Fibers with the Current Density Perceived from XPS Measurement (at Percent)

Current Density (A/m^2)	**Functional Groups**			
	CC (284.7 eV)	**COH (286.1 eV)**	**C=O (287.8 eV)**	**OHC=O (289.2 eV)**
1.6	51.2	29.1	10.2	6.6
0.8	50.1	28.9	10.0	9.6
0.4	53.0	28.3	9.1	6.8
0.2	54.1	28.2	9.0	6.7
0	63.3	25.4	4.2	2.7

It is observed in this work that the double bond in carboxyl and carbonyl functional groups on the surfaces of fiber can largely improve the physical pull at the interfaces between epoxy resins and fibers and, at the same time, the amide-kind functional groups. Thus, the growing amounts of oxygen-comprising functional groups on fibers play a vital role in enhancing the adhesion degree at interfaces between matrix and fibers. As noticed in Tables 5.6 and 5.7, it is also observed that larger upsurges of oxygen as compared to the nitrogen groups of fibers treated are associated with improving the bulk physical attraction degree as acidic functionality (Denison et al., 1986).

5.9.2. Contact Angle and the Surface Free Energy

Table 5.8 displays contact angle data of the carbon fibers without and with electrochemical oxidation. As an outcome, the water angle decreases largely with the growing current density equal to 0.2 to 0.4 A m^{-2}. This recommends that electrochemical oxidation triggers a change in the surface nature of the fiber, like hydrophobic-hydrophilic properties. In order to attain more detailed information regarding the surface energy of carbon fibers after and before electrochemical oxidation, an examination of the free energy of surface is assessed in surface energetic researches divided by two components: London dispersive component of the non-polar interaction and the specific component defining all other kinds of interactions (Keesom, Debye, and hydrogen bond). The specific and London dispersive components of the free energy of surfaces of the carbon fibers are concluded by determining the contact angles of the range of analysis liquids with well-known specific and London dispersive components and examining the outcomes in line with the approach proposed by Owens and Wendt (1969); and Kaelble (1970), utilizing the geometric mean approach.

Table 5.8. Determinations of Contact Angle Determinations on Carbon Fibers as the Function of Current Density

Wetting Liquids	**0 (A/m^2)**	**0.2 (A/m^2)**	**0.4 (A/m^2)**	**0.8 (A/m^2)**	**1.6 (A/m^2)**
Diiodomethane	36	29	26	25	29
Water	85	69	61	70	74

From the surface energetic viewpoint, it is anticipated that an upsurge in the specific (or polar) component of free energy of the surface of fibers will generally play a vital role in enhancing the adhesion degree at the interfaces between the epoxy resin matrix and the anodized fibers comprising oxide functional groups (Park et al., 1999; Marieta et al., 2005).

5.9.3. Mechanical Interfacial Properties

Fracture toughness is the critical property required to resist the propagation of crack loaded from the matrix to fiber, which must be well-thought-out in the assessment of the composite material for the real application. For the rectangular cross-section of composites, the fracture toughness of composites can usually be determined by the 3 points bending test for K_{IC} (critical stress intensity factor) (Table 5.9).

Table 5.9. Outcomes of the Free Energy of Surface and Their Constituents of Carbon Fibers as the Function of Current Density

Current Density (A/m^2)	**c^L_S [mJ/m^2]**	**cSPS [mJ/m^2]**	**c_S [mJ/m^2]**
1.6	42	5	47
0.8	44	6	50
0.4	42	11	53
0.2	42	7	49
0	40	2	42

Figure 5.10 exhibits the advancement of K_{IC} with the flexure of composites as the function of current density. In a K_{IC} fracture toughness experiment, the K_{IC} of composites recurrently upsurges with the growing current densities of treatments equal to 0.4 $A\ m^{-2}$, and the value of maximum strength is discovered around 283 Mega Pascals at an anodic treatment of

nearly 0.4 A m^{-2}. The extra energy required to spread the interfacial cracks under this circumstance is attributed to improved interfacial adhesion between matrix and fibers (Lindsay et al., 2007).

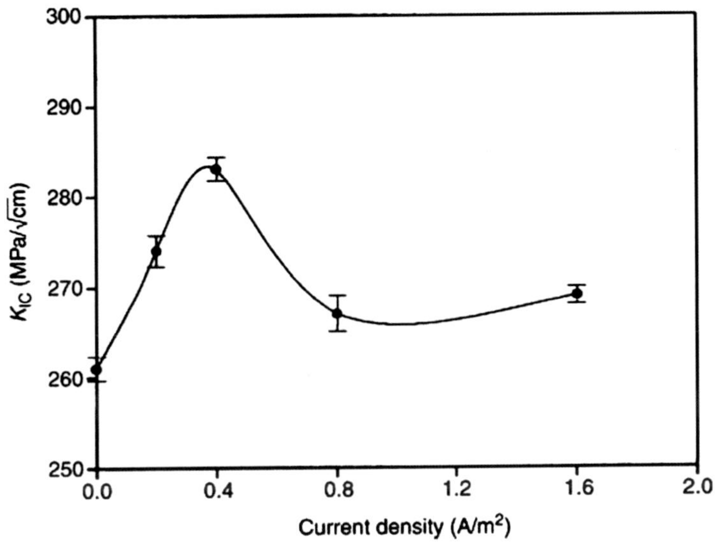

Figure 5.10. Evolution of K_{IC} as the function of current density planned.

Source: https://en.wikipedia.org/wiki/Fracture_toughness.

5.10. ELECTROPLATING EFFECT

5.10.1. Surface Characteristics

XPS spectra of the untreated and Ni-plated carbon fibers are displayed in Figure 5.12. As desired, the untreated carbon fibers (Figure 5.12(a)) exhibit the O_{1s} peak and the significant C_{1s} peak at 532.8 and 284.6 eV, correspondingly. The peak of O_{1s} is probably because of the intrinsic surface carboxyl or carbonyl groups. Then, for the Ni-plated carbon fibers (Figure 5.12(b–e)), nickel, oxygen, and carbon peaks are noticed in the XPS spectra. The O_{1s} peak of the Ni-plated carbon fibers is possible because of the C=O, NiO, OCO, and OH groups on the surface of the fiber (Figure 5.11) (Yang et al., 2008).

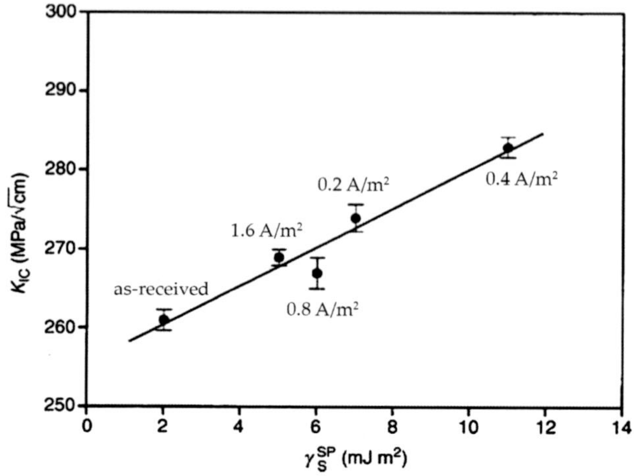

Figure 5.11. Dependence of the K_{IC} of composites on a specific (or polar) component of a fiber surface free energy planned.

Source: https://analyticalsciencejournals.onlinelibrary.wiley.com/doi/abs/10.1002/sia.740030506.

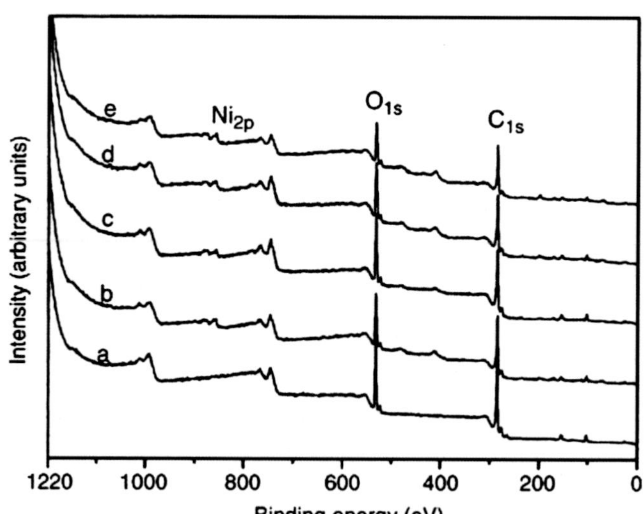

Figure 5.12. *XPS spectra of untreated and Ni-plated carbon fibers: (a) 0 A m^{-2}; (b) 5 A m^{-2}; (c) 10 A m^{-2}; (d) 30 A m^{-2}; and (e) 60 A m^{-2}.*

Source: https://www.sciencedirect.com/science/article/abs/pii/S0008622300001846.

Figure 5.13(a) displays the expanded scale of O_{1s} XPS spectra for the carbon fibers covered with Ni at 10 A m^{-2} current density. The spectra of O_{1s} reveal the existence of 3 peaks conforming to NiO groups (BE=529.6 eV), OH or C=O groups (BE=531.6 eV), and the OCO groups (BE=532.6 eV).

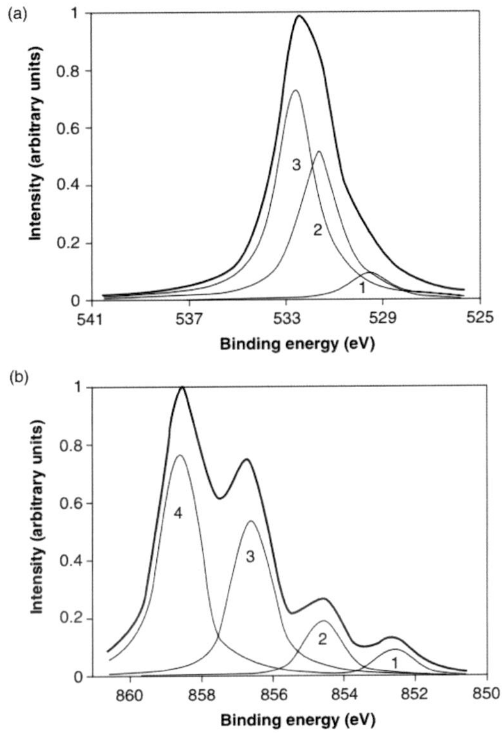

Figure 5.13. High-resolution Ni_{2p} and O_{1s} XPS spectra of the Ni-plated carbon fibers (10 A m^{-2} current density). (a) NiO (peak 1, BE= 529.6 eV), C=O or OH (peak 2, BE= 531.6 eV), OCO (peak 3, BE= 532.6 eV; (b) Ni-metal (peak 1, BE= 852.7 eV), NiO (peak 2, BE= 854.6 eV), Ni(OH)2 (peak 3, BE= 856.4 eV), and Ni-metal (peak 4, BE= 858.5 eV).

Source: https://www.sciencedirect.com/science/article/abs/pii/S0169433209005297.

The peak of Ni_{2p} is displayed on the expanded scale in Figure 5.13(b) for Ni-plated carbon fibers. In the situation of Ni-plated carbon fibers in Figure 5.13(b), various sub-peaks are observed along the key peak (BE=858.5 eV). The sub-peaks that are computationally separated in Figure 5.13(b) specify that nickel metal (BE=852.7 eV, and BE=858.5 eV), NiO (BE=854.6 eV), and $Ni(OH)_2$ (BE=856.4 eV) are existent as an outcome of electrolytic Ni plating. Conversely, the peaks are never observed for untreated carbon fibers. For the nickel-plated and untreated carbon fibers, the elemental composition, and the ratio of O_{1s}/C_{1s} are given in Table 5.10. The C_{1s}/O_{1s} composition ratios of the Ni-plated carbon fibers are augmented than that of the untreated because of the deposition of active forms, like NiO, Ni metal, and $Ni(OH)_2$ on an inactive carbon.

Table 5.10. O_{1s}/C_{1s} Ratio and Elemental Composition of the Ni-Plated and Untreated Carbon Fibers (at Percent)

Current Density (A/m^2)	XPS			
	C_{1s}	O_{1s}	N_{1s}	O_{1s}/C_{1s}
60	63.0	23.9	0.8	0.379
30	63.3	26.8	0.8	0.423
10	62.8	28.0	0.8	0.446
5	64.4	25.4	0.8	0.394
0	68.8	25.8	0.8	0.375

From the XPS investigational outcomes, it is discovered that the surface composition of the carbon fibers varied substantially as an outcome of Ni plating. The carbon (C) content of the Ni-plated fibers was reduced when these fibers were plated with the metallic Ni, while the nickel and oxygen contents of the plated fibers were much higher as compared to that for the untreated fibers. The vigorous groups on the surfaces of carbon fiber after Ni plating might help to vary the functionality and polarity of the surfaces of fiber (Park and Jang, 2001; Costa et al., 2005).

5.10.2. Contact Angle and the Surface Free Energy

Contact angles of the Ni-plated and untreated carbon fibers were normally measured for two testing liquids, diiodomethane, and deionized water. The γ^L_S, $\gamma^{SP}_{S,}$ and γ_S of carbon fibers researched are given in Figure 5.14. The treatments at a current density at 5 to 60 A m^{-2} triggered a wetting level much better as compared to those for the specimen without plating. The c^{SP}_{S} (polar component) of the Ni-plated carbon fiber was considerably augmented, but the c^L_S (dispersive component) was hardly altered. In this kind of system, the carbon fibers' surface polarity is also increased than the untreated ones. These outcomes disclose that electroplating triggers an enhancement of interaction amongst the carbon fiber surfaces and the polar liquid (Yang et al., 1994; Gardner et al., 1995).

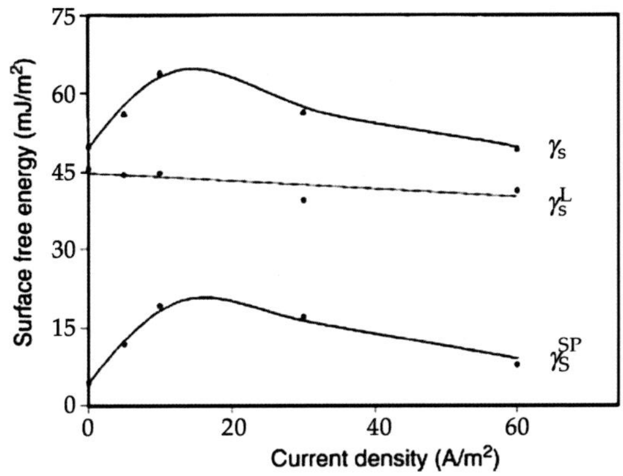

Figure 5.14. Free energy of surface the Ni-plated and untreated carbon fibers [gS: surface free energy; gLS: London dispersive component; and gSPS: specific component].

Source: https://www.researchgate.net/figure/Surface-free-energy-and-their-components-of-carbon-fibers_fig2_264013017.

To enhance wettability, the surface free energy of carbon fibers must be made equal to or larger than the surface energy of a matrix. In the circumstance of 10 A m^{-2} current density, the free energy of the surface reaches the maximum value of 67 mJ m^{-2}.

5.10.3. Mechanical Interfacial Properties

An analytical expression for K_C in the mode I fracture (K_{IC}) is characterized utilizing the SEN beam fracture test of toughness in 3 points bending flexure. Figure 5.15 exhibits K_{IC} of the nickel-plated carbon fiber-strengthened composites as the function of electric current density. The utmost strength is discovered at the current density of nearly 10 A m^{-2}. Also, it is noticed that the K_{IC} of composites for fibers handled at a comparatively high current density (60 A m^{-2}) is not considerably increased than that of composites prepared with untreated fibers. However, in this kind of system, it seems that Ni plating of the surfaces of carbon fiber affects the K_{IC} of composites, consequential from the existence of nickeled functional groups and augmented oxide functional groups on the surface of carbon fibers (Brown et al., 1992).

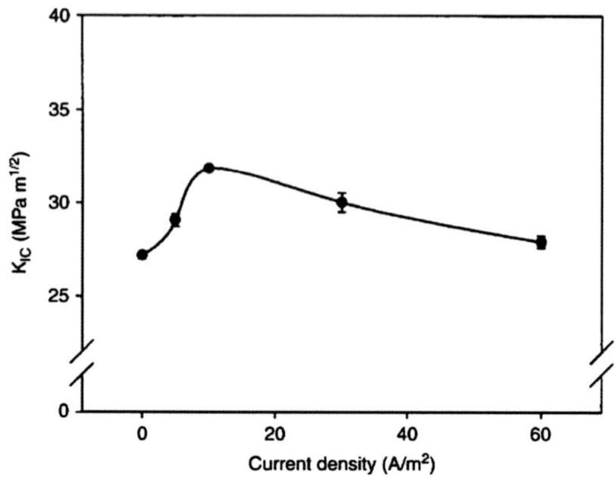

Figure 5.15. KIC of nickel-plated carbon fiber-strengthened composites with diverse current densities.

Source: https://onlinelibrary.wiley.com/doi/book/10.1002/9783527645213.

It is very interesting to observe that the K_{IC} outcomes support the dependability of data, as the trends in the values of K_{IC} values appear to be very analogous to trends in surface free energy or O_{1s}/C_{1s}. Actually, the associations between O_{1s}/C_{1s} and K_{IC} and between $c^{SP}{}_{S}$ and K_{IC} are nearly linear for all of the samples with diverse current densities, as displayed in Figures 5.16 and 5.17, correspondingly. Thus, the $c^{SP}{}_{S}$ and O_{1s}/C_{1s} ratio might demonstrate (Figure 5.18)

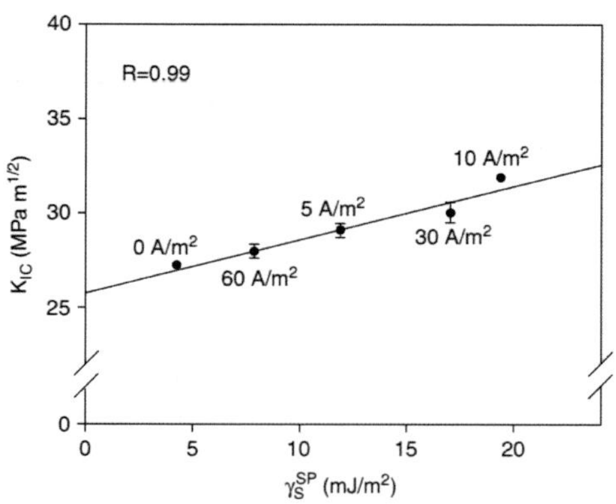

Figure 5.16. *Dependence of K_{IC} on the O_{1s}/C_{1s} ratio (where R → coefficient of regression).*

Source: https://www.wiley.com/en-us/Polymer+Composites%2C+Volume+1%2C+Macro+and+Microcomposites-p-9783527326242.

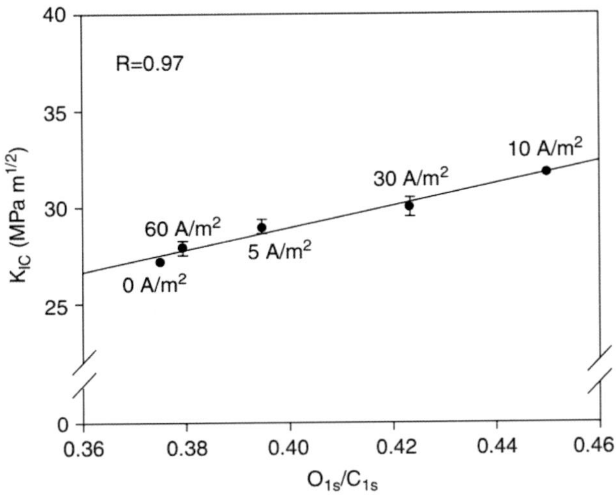

Figure 5.17. Dependence of K_{IC} on the specific polar component g^{SP}_{S} of the free energy of surface (where R → coefficient of regression).

Source: https://www.wiley.com/en-us/Polymer+Composites%2C+Volume+1%2C+Macro+and+Microcomposites-p-9783527326242.

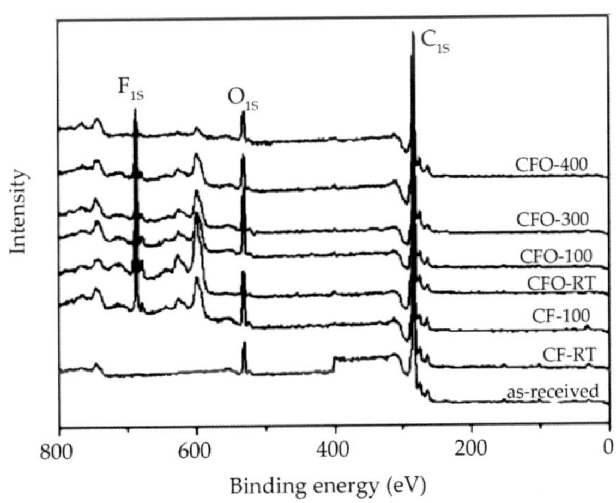

Figure 5.18. XPS broad scan spectra of oxyfluorinated carbon fibers.

Source: https://www.semanticscholar.org/paper/Influence-of-Oxyfluorination-on-Properties-of-(PAN)-Lim-Yoo/2652ce1e026fd6f84f9b5476f6e5a9907e63cb9f/figure/1.

To be the controlling factors in adhesion between the matrix resins and the fibers in the existent system studied (Chung, 2001).

5.11. APPLICATIONS

After World War II, US manufacturers started producing polyester resin and fiberglass composite radomes, and boat hulls. The automotive industry 1st introduced composites into vehicle bodies in the early 1950s. Due to the highly anticipated corrosion resistance, high strength, and lightweight characteristics in the composites, research focus went into enhancing the manufacturing process and material science. Such struggles led to the growth of two novel manufacturing methods known as pultrusion and filament winding, which assisted the progress of the composite technology into the novel markets. There was an increased demand by the industry of recreation for tennis rackets, composite fishing rods, golf clubs, and ski equipment. The aerospace industry started to utilize composites in containers, pressure vessels, and non-structural aircraft components. The customers also started installing composite covers, bathtubs, ladders, railings, and electrical equipment. The 1st civil use in composites was the dome structure made 1968 in Benghazi in 1968, and the other structures trailed slowly (Podgoric et al., 2009).

Through DOE sponsorship, ORNL (Oak Ridge National Laboratory) and the partners have normally been functioning with Automotive Composites Consortium in order to mature technologies that would empower the making of commercial-grade carbon fiber at nearly $5/lb to $7/lb. The accomplishment of this price goal would permit the institution of carbon fiber-centered composites into numerous applications for future vehicles and speed up its implantation in the other energy-associated applications. The objective of lower price carbon fiber-strengthened composites has compelled the development of alternate precursors and more effective production methods, along with novel composites manufacturing methods.

5.11.1. Automotives

Carbon fiber-strengthened polymer is utilized broadly in high-end automobile racing. The expense of carbon fiber is lessened by the unmatched strength to weight ratio of the material, and the low weight is necessary for high-performance automobile racing. The manufacturers of race cars have also developed approaches to give pieces of carbon fiber strength in a particular direction, making it quite strong in the direction of load-bearing, but weak in the directions where no or little load would generally be placed on a member (Tang et al., 2008).

5.11.2. Wind Energy

Judicious utilization of carbon fibers decreases the blade weight utilized in large wind turbines. Larger blades upsurge energy capture and reduce the electricity cost. Therefore, turbines have generally been rising larger in a push to decrease the energy cost produced from the renewable sources of wind to make the price of wind-produced energy economical with the expense of energy produced from the conventional power sources.

5.11.3. Deepwater Offshore Gas and Oil Production

The expense of the floating platforms can generally be considerably decreased by decreasing their size. Decreasing the weight of several components-particularly umbilicals, riser strings, and the other structures adjourned from the platforms-can usually be achieved utilizing carbon fibers, therefore permitting platform shrinkage, and decreasing the expense of deep-water gas and oil production.

5.11.4. Electricity Transmission

Carbon fiber composite-covered transmission conductors can normally increase (double) the current-carrying capability of distribution and transmission power cables. Low thermal expansion of carbon fibers also virtually removes the elevated temperature line sag, therefore increasing the reliability of the system.

5.11.5. Commercial Aircraft

The composites of carbon fiber have long been utilized in aerospace and aircraft applications, but the broad utilization of carbon fibers in the next generation commercial aircraft is stated to decrease fuel demand by nearly 20% as well as provides benefits that eventually deliver enhanced passenger comfort. Most of the Airbus A350 XWB's and new Boeing 787 Dreamliner's fuselage will be made up of CFRP, thus making the aircraft much lighter as compared to the aluminum fuselage, with the extra advantage of less upkeep, thanks to the superior fatigue resistance of CFPR.

5.11.6. Civil Infrastructure

Even though composites have usually been widely developed and utilized in the marine and aerospace industries, these technologies do not meet the particular requirements of civil applications. Thus, composites are gradually gaining recognition in the civil and construction industry. Composites are utilized for upgrading the bridge columns in order to fulfill seismic standards, along with patching cracks in masonry and concrete structures.

REFERENCES

1. Ahearn, C., & Rand, B., (1996). Modification of the fiber-matrix bonding in a brittle carbon-carbon composite by controlled oxidation. *Carbon, 34*(2), 239–249.
2. Beydon, R., Bernhart, G., & Segui, Y., (2000). Measurement of metallic coatings adhesion to fiber-reinforced plastic materials. *Surface and Coatings Technology, 126*(1), 39–47.
3. Brelant, S., (1981). Chemical interaction between the carbon fiber surface and epoxy resins. *Carbon, 19*(4), 329.
4. Brown, N. M., Hewitt, J. A., & Meenan, B. J., (1992). X-ray-induced beam damage observed during x-ray photoelectron spectroscopy (XPS) studies of palladium electrode ink materials. *Surface and Interface Analysis, 18*(3), 187–198.
5. Chong, Y. B., & Ohara, H., (1992). Modification of carbon fiber surfaces by direct fluorination. *Journal of Fluorine Chemistry, 57*(1–3), 169–175.
6. Chung, D. D. L., (2001). Electromagnetic interference shielding effectiveness of carbon materials. *Carbon, 39*(2), 279–285.
7. Chwastiak, S., (1973). A wicking method for measuring wetting properties of carbon yarns. *Journal of Colloid and Interface Science, 42*(2), 298–309.
8. Clayton, A., (1988). *Epoxy Resins: Chemistry and Technology* (Vol. 1, pp. 1–26). M. Dekker.
9. Costa, J., Turon, A., Trias, D., Blanco, N., & Mayugo, J. A., (2005). A progressive damage model for unidirectional fiber-reinforced composites based on fiber fragmentation. Part II: Stiffness reduction in environment sensitive fibers under fatigue. *Composites Science and Technology, 65*(14), 2269–2275.
10. Curran, K., Underhill, M., Gibson, L. T., & Strlic, M., (2016). The development of a SPME-GC/MS method for the analysis of VOC emissions from historic plastic and rubber materials. *Microchemical Journal, 124*, 909–918.
11. Delamar, M., Desarmot, G., Fagebaume, O., Hitmi, R., Pinsonc, J., & Savéant, J. M., (1997). Modification of carbon fiber surfaces by electrochemical reduction of aryl diazonium salts: Application to carbon epoxy composites. *Carbon, 35*(6), 801–807.

12. Denison, P., Jones, F. R., & Watts, J. F., (1986). XPS analysis of barium labeled carbon-fibers and interfacial effects in fiber epoxy composites. *Surface and Interface Analysis, 9*(1–6), 431–435.
13. Denison, P., Jones, F. R., & Watts, J. F., (1987). The use of XPS and labelling techniques to study the surface chemistry of carbon fibers. *Journal of Physics D: Applied Physics, 20*(3), 306.
14. Drzal, L. T., Rich, M. J., & Lloyd, P. F., (1983). Adhesion of graphite fibers to epoxy matrices: I. The role of fiber surface treatment. *The Journal of Adhesion, 16*(1), 1–30.
15. Fukunaga, A., & Ueda, S., (2000). Anodic surface oxidation for pitch-based carbon fibers and the interfacial bond strengths in epoxy matrices. *Composites Science and Technology, 60*(2), 249–254.
16. Gardner, S. D., Singamsetty, C. S., Booth, G. L., He, G. R., & Pittman, Jr. C. U., (1995). Surface characterization of carbon fibers using angle-resolved XPS and ISS. *Carbon, 33*(5), 587–595.
17. Gopal, P., Dharani, L. R., & Blum, F. D., (1997). *Fracture Behavior of Chopped Glass Strand Reinforced Phenolic Composites, 1,* 1–23.
18. Hage, E., Costa, S. F., & Pessan, L. A., (1997). Modification of the carbon fiber surface with a copper coating for composite materials with epoxy. *Journal of Adhesion Science and Technology, 11*(12), 1491–1499.
19. Ho, K. K., Lamoriniere, S., Kalinka, G., Schulz, E., & Bismarck, A., (2007). Interfacial behavior between atmospheric-plasma-fluorinated carbon fibers and poly(vinylidene fluoride). *Journal of Colloid and Interface Science, 313*(2), 476–484.
20. Hoecker, F., & Karger-Kocsis, J., (1993). Effects of crystallinity and supermolecular formations on the interfacial shear strength and adhesion in GF/PP composites. *Polymer Bulletin, 31*(6), 707–714.
21. Jensen, C., Zhang, C., & Qiu, Y., (2003). The aging of atmospheric plasma-treated ultrahigh-modulus polyethylene fibers. *Composite Interfaces, 10*(2, 3), 277–285.
22. Jones, C., (1991). The chemistry of carbon fiber surfaces and its effect on interfacial phenomena in fiber/epoxy composites. *Composites Science and Technology, 42*(1–3), 275–298.
23. Jones, C., (1993). Effects of electrochemical and plasma treatments on carbon fiber surfaces. *Surface and Interface Analysis, 20*(5), 357–367.

24. Kaelble, D. H., (1970). Dispersion-polar surface tension properties of organic solids. *The Journal of Adhesion, 2*(2), 66–81.
25. Kawai, M., Yajima, S., Hachinohe, A., & Kawase, Y., (2001). High-temperature off-axis fatigue behavior of unidirectional carbon-fiber-reinforced composites with different resin matrices. *Composites Science and Technology, 61*(9), 1285–1302.
26. Kennedy, J. H., & Frese, Jr. K. W., (1978). Photooxidation of water at α-Fe_2O_3 electrodes. *Journal of the Electrochemical Society, 125*(5), 709.
27. Kinloch, A. J., (1987). *Adhesion and Adhesives Chapman and Hall, 1*, 1–20. New York.
28. Kozlowski, C., & Sherwood, P. M., (1985). X-ray photoelectron-spectroscopic studies of carbon-fiber surfaces; Part 5. The effect of pH on surface oxidation. *Journal of the Chemical Society, Faraday Transactions 1: Physical Chemistry in Condensed Phases, 81*(11), 2745–2756.
29. Kozlowski, C., & Sherwood, P. M., (1986). X-ray photoelectron spectroscopic studies of carbon fiber surfaces vii-electrochemical treatment in ammonium salt electrolytes. *Carbon, 24*(3), 357–363.
30. Kumar, S., Rath, T., Mahaling, R. N., Reddy, C. S., Das, C. K., Pandey, K. N., & Yadaw, S. B., (2007). Study on mechanical, morphological, and electrical properties of carbon nanofiber/polyetherimide composites. *Materials Science and Engineering: B, 141*(1, 2), 61–70.
31. Lawton, R. A., Price, C. R., Runge, A. F., Doherty, III. W. J., & Saavedra, S. S., (2005). Air plasma treatment of submicron thick PDMS polymer films: Effect of oxidation time and storage conditions. *Colloids and Surfaces A: Physicochemical and Engineering Aspects, 253*(1–3), 213–215.
32. Lee, W. S., Sue, W. C., & Lin, C. F., (2000). The effects of temperature and strain rate on the properties of carbon-fiber-reinforced 7075 aluminum alloy metal-matrix composite. *Composites Science and Technology, 60*(10), 1975–1983.
33. Li, H., Liang, H., He, F., Huang, Y., & Wan, Y., (2009). Air dielectric barrier discharges plasma surface treatment of three-dimensional braided carbon fiber reinforced epoxy composites. *Surface and Coatings Technology, 203*(10, 11), 1317–1321.

34. Lindsay, B., Abel, M. L., & Watts, J. F., (2007). A study of electrochemically treated PAN based carbon fibers by IGC and XPS. *Carbon, 45*(12), 2433–2444.
35. Ling, T., Lu, Y., Zhang, Z., Li, C., & Oeser, M., (2019). Value-added application of waste rubber and waste plastic in asphalt binder as a multifunctional additive. *Materials, 12*(8), 1280.
36. Mallick, P. K., (2007). *Fiber-Reinforced Composites: Materials, Manufacturing, and Design, 1*, 1–27. CRC press.
37. Marieta, C., Schulz, E., Irusta, L., Gabilondo, N., Tercjak, A., & Mondragon, I., (2005). Evaluation of fiber surface treatment and toughening of thermoset matrix on the interfacial behavior of carbon fiber-reinforced cyanate matrix composites. *Composites Science and Technology, 65*(14), 2189–2197.
38. Meservey, R., & Tedrow, P. M., (1978). Surface relaxation times of conduction-electron spins in superconductors and normal metals. *Physical Review Letters, 41*(12), 805.
39. Michalski, M. C., & Saramago, B. J., (2000). Static and dynamic wetting behavior of triglycerides on solid surfaces. *Journal of Colloid and Interface Science, 227*(2), 380–389.
40. Miyahira, A. K., & Soule, H. R., (2019). The 25th annual prostate cancer foundation scientific retreat report. *The Prostate, 79*(12), 1363–1378.
41. Montes-Morán, M. A., Van, H. F. W. J., Nunes, J. P., Martínez-Alonso, A., Tascón, J. M. D., & Bernardo, C. A., (2005). A study of the effect of plasma treatment on the interfacial properties of carbon fiber-thermoplastic composites. *Carbon, 8*(43), 1795–1799.
42. Morgan, P., (2005). *Carbon Fibers and Their Composites, 1*, 1–23. CRC press.
43. Morra, M., Occhiello, E., Garbassi, F., & Nicolais, L., (1991). Surface studies on untreated and plasma-treated carbon fibers. *Composites Science and Technology, 42*(4), 361–372.
44. Nakajima, T., (2000). Fluorine-containing energy conversion materials. *Journal of Fluorine Chemistry, 105*(2), 229–238.
45. Nakao, F., Takenaka, Y., & Asai, H., (1992). Surface characterization of carbon fibers and interfacial properties of carbon fiber composites. *Composites, 23*(5), 365–372.
46. Neumann, A. W., David, R., & Zuo, Y., (2010). *Applied Surface Thermodynamics, 151*, 1–20. CRC press.

47. Oliver, S. R., Bowden, N., & Whitesides, G. M., (2000). Self-assembly of hexagonal rod arrays based on capillary forces. *Journal of Colloid and Interface Science, 224*(2), 425–428.

48. Owens, D. K., & Wendt, R. C., (1969). Estimation of the surface free energy of polymers. *Journal of Applied Polymer Science, 13*(8), 1741–1747.

49. Paiva, M. C., Bernardo, C. A., & Nardin, M., (2000). Mechanical, surface, and interfacial characterization of pitch and PAN-based carbon fibers. *Carbon, 38*(9), 1323–1337.

50. Park, S. J., & Jang, Y. S., (2001). Interfacial characteristics and fracture toughness of electrolytically Ni-plated carbon-fiber-reinforced phenolic resin matrix composites. *Journal of Colloid and Interface Science, 237*(1), 91–97.

51. Park, S. J., & Kim, B. J., (2005). Roles of acidic functional groups of carbon fiber surfaces in enhancing interfacial adhesion behavior. *Materials Science and Engineering: A, 408*(1, 2), 269–273.

52. Park, S. J., & Kim, J. S., (2001). Modifications produced by electrochemical treatments on carbon blacks: Microstructures and mechanical interfacial properties. *Carbon, 39*(13), 2011–2016.

53. Park, S. J., & Kim, M. H., (2000). Effect of acidic anode treatment on carbon fibers for increasing fiber-matrix adhesion and its relationship to interlaminar shear strength of composites. *Journal of Materials Science, 35*(8), 1901–1905.

54. Park, S. J., & Lee, J. R., (1998). Bending fracture and acoustic emission studies on carbon-carbon composites: Effect of sizing treatment on carbon fibers. *Journal of Materials Science, 33*(3), 647–651.

55. Park, S. J., & Park, B. J., (1999). Electrochemically modified PAN carbon fibers and interfacial adhesion in epoxy-resin composites. *Journal of Materials Science Letters, 18*(1), 47–49.

56. Park, S. J., (1999). *Interfacial Forces and Fields: Theory and Applications* (Vol. 1, pp. 380–385). Marcel Dekker, New York.

57. Park, S. J., Jin, J. S., & Lee, J. R., (2000). Influence of silane coupling agents on the surface energetics of glass fibers and mechanical interfacial properties of glass fiber-reinforced composites. *Journal of Adhesion Science and Technology, 14*(13), 1677–1689.

58. Park, S. J., Park, W. B., & Lee, J. R., (1999). Roles of unsaturated polyester in the epoxy matrix system. *Polymer Journal, 31*(1), 28–31.

59. Park, S. J., Seo, M. K., & Lee, J. R., (2003b). Relationship between surface characteristics and interlaminar shear strength of oxyfluorinated carbon fibers in a composite system. *Journal of Colloid and Interface Science, 268*(1), 127–132.
60. Park, S. J., Seo, M. K., & Rhee, K. Y., (2003a). Studies on mechanical interfacial properties of oxy-fluorinated carbon fiber-reinforced composites. *Materials Science and Engineering: A, 356*(1, 2), 219–226.
61. Park, S. J., Seo, M. K., Ma, T. J., & Lee, D. R., (2002). Effect of chemical treatment of Kevlar fibers on mechanical interfacial properties of composites. *Journal of Colloid and Interface Science, 252*(1), 249–255.
62. Pittman, Jr. C. U., Jiang, W., He, G. R., & Gardner, S. D., (1998). Oxygen plasma and isobutylene plasma treatments of carbon fibers: Determination of surface functionality and effects on composite properties. *Carbon, 36*(1, 2), 25–37.
63. Podgoric, S., Jones, B. J., Bulpett, R., Troisi, G., & Franks, J., (2009). Diamond-like carbon/epoxy low-friction coatings to replace electroplated chromium. *Wear, 267*(5–8), 996–1001.
64. Proctor, A., & Sherwood, P. M., (1982). X-ray photoelectron spectroscopic studies of carbon fiber surfaces: I. carbon fiber spectra and the effects of heat treatment. *Journal of Electron Spectroscopy and Related Phenomena, 27*(1), 39–56.
65. Proctor, A., & Sherwood, P. M., (1982). X-ray photoelectron spectroscopic studies of carbon fiber surfaces. III: Industrially treated fibers and the effect of heat and exposure to oxygen. *Surface and Interface Analysis, 4*(5), 212–219.
66. Rapley, J. W., Swan, R. H., Hallmon, W. W., & Mills, M. P., (1990). The surface characteristics produced by various oral hygiene instruments and materials on titanium implant abutments. *International Journal of Oral & Maxillofacial Implants, 5*(1), 1–23.
67. Ryu, S. K., Park, B. J., & Park, S. J., (1999). XPS analysis of carbon fiber surfaces—anodized and interfacial effects in fiber-epoxy composites. *Journal of Colloid and Interface Science, 215*(1), 167–169.
68. Schwartz, M. M., (1983). *Composite Materials Handbook, 1*, 1–20.

69. Seo, M. K., Park, S. J., & Lee, S. K., (2005). Influence of atmospheric plasma on physicochemical properties of vapor-grown graphite nanofibers. *Journal of Colloid and Interface Science, 285*(1), 306–313.
70. Severini, F., Formaro, L., Pegoraro, M., & Posca, L., (2002). Chemical modification of carbon fiber surfaces. *Carbon, 40*(5), 735–741.
71. Tang, Y., Liu, H., Zhao, H., Liu, L., & Wu, Y., (2008). Friction and wear properties of copper matrix composites reinforced with short carbon fibers. *Materials & Design, 29*(1), 257–261.
72. Wang, S., & Garton, A., (1992). Chemical interactions at the interface between a carbon fiber and a boron trifluoride-catalyzed epoxy matrix. *Journal of Applied Polymer Science, 45*(10), 1743–1752.
73. Wang, T., (1995). *Chemical Modification of Interface of Carbon Fiber Reinforced Carbon Composites and Carbon Fiber Reinforced Polymers, 1*, 1–20.
74. Wu, S., (1982). *Polymer Interface and Adhesion* (Vol. 1, pp. 1–32). Marcel Dekker. Inc., Wilmington, Delaware.
75. Wu, Z., Pittman, Jr. C. U., & Gardner, S. D., (1995). Nitric acid oxidation of carbon fibers and the effects of subsequent treatment in refluxing aqueous NaOH. *Carbon, 33*(5), 597–605.
76. Yang, F., Zhang, X., Han, J., & Du, S., (2008). Mechanical properties of short carbon fiber reinforced ZrB_2-SiC ceramic matrix composites. *Materials Letters, 62*(17, 18), 2925–2927.
77. Yang, W. P., Costa, D., & Marcus, P., (1994). Resistance to pitting and chemical composition of passive films of a Fe-17% Cr alloy in chloride-containing acid solution. *Journal of the Electrochemical Society, 141*(10), 2669.
78. Yosomiya, R., (2020). *Adhesion and Bonding in Composites, 1*, 1–20. CRC Press.

Chapter

6

Kevlar Fiber-Reinforced Polymer Composites

CONTENTS

6.1. Introduction 198

6.2. Kevlar Fiber 199

6.3. Interface 201

6.4. Factors Affecting the Composite Properties 202

6.5. Surface Modification 207

References 211

6.1. INTRODUCTION

In various application fields, thermoplastic-based composites are becoming more significant because of the association of toughness of thermoplastic polymers together with the strength and stiffness of reinforcing fibers. Polymeric composites can be prepared by using Organic textile fibers. For both thermoplastics and rubbers, Organic textile fibers are utilized being a reinforcing agent in polymer matrices because of their low stiffness. The nature of the polymeric material is remarkably strongly dependent on its morphology, structure, and relaxation processes (Figure 6.1) (Kadolph, 2002; Blicblau, 2005).

Figure 6.1. Applications of aramid fiber.

Source: https://global.ipitaka.com/blogs/news/what-are-aramid-fibers-i?region_redirected=true.

Moreover, the properties of the polymer matrices themselves determine the characteristics of composite materials, along with reinforcements, and also the adhesion of fiber/matrix interface, which depends chiefly on the bonding strength at the interface and the voids. Generally, the Kevlar fibers are strongly crystalline, and their surface is smooth and chemically inert; therefore, it has too much poor adhesion with the matrix. Hence, for the improvement of the reinforcing efficiency of the Kevlar fiber, its surface modification is essential. For the surface modification of Kevlar fibers, an effective approach is the direct fluorination. Any catalyst or initiation is not required by this process (Peters, 2013). The surface of the fibers is modified using elemental gases. The fluorination can be executed at room temperature.

Incorporation in the polymer matrix and the surface modification of Kevlar fibers through fluorination turned to better thermostability, mechanical stability, and membrane properties (Mukherjee, 2007).

6.2. KEVLAR FIBER

Kevlar refers to the DuPont trade name of poly(*p*-phenylene terephthalamide) (PPTA). Initially, it was created in 1965 at DuPont's labs by some scientists. In the aromatic polyamide (PA) family, it is an organic fiber. It has a unique combination of high modulus, high strength, thermal stability, and toughness. It can be spun into sheets or ropes of fabric that can be utilized in the manufacturing of composite components. Kevlar is utilized in a vast range of applications from bicycles to body armor because of its high strength-to-weight ratio. Moreover, in contrast to steel, it is five times stronger on an equal weight basis. It is a competitor of Twaron created by Teijin, and a member of the Aramid family of synthetic fibers (Clegg and Collyer, 1986).

6.2.1. Molecular Structure and Development of Kevlar

A new technique has been discovered by a company in 1965 for making perfect polymer chain extension. Liquid crystalline solutions are found to be formed by the polymer poly*p*-benzamide because its molecular backbone is simply repetitive. The *para* orientation on the benzene ring is the key structural need for the backbone, which makes it able to formulate rodlike molecular structures. These developments result in the latest formulation of Kevlar. The fiber of poly(*p*-phenylene terepthalamide) is being developed by the same company, which was once known to be a high strength Kevlar aramid fiber, in 1971. The fiber-forming substance in aramid fiber is a long-chain synthetic PA where a minimum of 85% of the amide linkages is connected directly to two aromatic rings. Hence, most of the amide groups in an aramid are directly attached to two aromatic rings, just intervening (Figure 6.2) (Sadov et al., 1978).

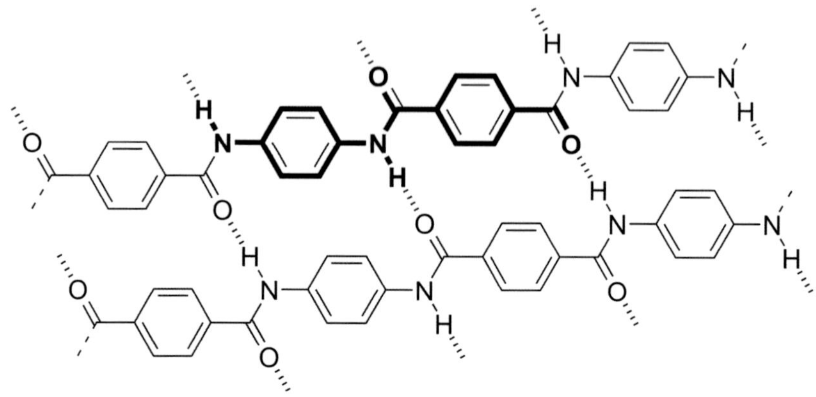

Figure 6.2. Molecular structure of Kevlar and hydrogen bonding.

Source: https://en.wikipedia.org/wiki/Kevlar.

Based on the chemical and physical microstructures of the Kevlar fibers, its properties can be explained. While the process of spinning, the polymer chains are oriented in the draw direction forming chain crystallites possessing a very high aspect ratio. Hydrogen bonding connects the chains within the crystallites that makes these fibers highly strong. If the hydrogen bonds are broken at once, then these fibers can break by tension. This is the only way, but it is very difficult to execute due to their huge numbers (Morley, 1976).

6.2.2. Properties of Kevlar Fibers

- These fibers are flame resistant; still they can be ignited.
- Aramids possess very high thermal stability due to their ordered and highly aromatic structure.
- Among all composite materials, Kevlar fiber composites possess the highest specific strengths.
- In contrast to carbon dispersed glass fibers, Kevlar fiber possesses low electrical conductivity and chemical inertness.
- Aramids propose remarkable combinations of properties for certain applications, e.g., high specific strength, creep resistance, toughness, and moderate cost, even though composites from newer fibers have enhanced that position (Table 6.1).

6.3. INTERFACE

A significant role is offered by the properties and structure of the fiber/matrix interface in administrating the physical and mechanical properties of the fiber-reinforced composites.

Table 6.1. Unique Properties of Kevlar Fiber in Comparison with Other Fibers

Properties	Specific Gravity	Tensile Strength (MPa)	Tensile Modulus (GPa)	Elongation at Break (%)
Nylon 66	1.14	1000	5.52	18
Kevlar 29	1.44	2750	82.7	5.2
Kevlar 49	1.45	2760	131	2.4
E-Glass	2.55	1700	68.9	3.0
Steel	7.86	1960	200	2.4

In fiber-reinforced thermoplastics, the surface of the matrix, the surface of the fiber side, and the phase between matrix and fiber are collectively known as interface. The enormous difference between the properties of the matrix and the fibers are reached out by the interface, or in other words, the interface makes it able to transfer the stress acting on the matrix to the fibers (the principal load-bearing constituents). These interface effects find out to be a kind of adhesion phenomenon and are often analyzed in terms of the surface structure of the bonded materials, i.e., surface factors like surface free energy, wettability, the surface roughness of the material to be bonded, and the presence of the polar groups on the surface. For the theoretical interpretation of the stress transfer from the fiber to the matrix, many assumptions about the properties of the interface must be made that are as follows (Akay and Barkley, 1991; Fu et al., 1999):

- The interfacial bond is infinitesimally tenuous;
- The fibers and the matrix act as classic materials;
- The bond between the matrix and the fiber is perfect;
- The materials near the fiber possess the same properties being the material in bulk form;
- The fibers are organized in a repeating or regular array.

In polymeric materials, some mathematical models are essential to correlate such assumptions. The fracture surfaces of fiber-reinforced composite having various modes of the interface are illustrated in Figure 6.3.

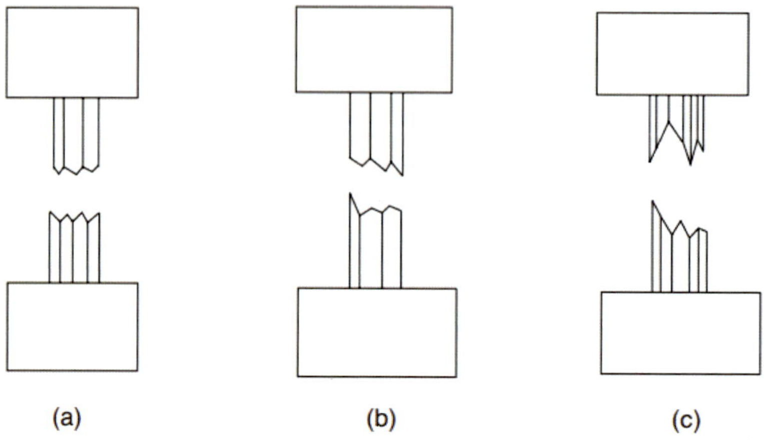

Figure 6.3. Illustrations of a classic fracture surface of unidirectional composite loaded in tension with the fiber direction: (a) composite for strong interfacial bond; (b) composite for an intermediate interfacial bond; and (c) composite for a poor interfacial bond.

Source: https://www.researchgate.net/publication/230249040_Fluorinated_and_oxyfluorinated_short_Kevlar_fiber-reinforced_ethylene_propylene_polymer.

6.4. FACTORS AFFECTING THE COMPOSITE PROPERTIES

6.4.1. Modulus, Strength, and Chemical Stability of the Polymer Matrix and the Fiber

The modulus and strength of the reinforcements broadly regulate the mechanical characteristics of fiber-reinforced composites. As it is discussed earlier, those fibers are the main load-bearing constituents in fiber-reinforced composites, whereas the matrix maintains those fibers in their desired orientations and positions. The matrix must be selected as per the needs of the end product. There are several other factors that should be considered also like the ease of fabrication, cost, and environmental conditions (e.g., humidity, temperature) in which the end product is going to be applied. While regulating the composite properties, a supreme significance is also deserved by the required durability of the end product, the weathering and chemical resistance of the matrix, and design stresses. In a fiber-reinforced composite, the main role of the matrix will change based on the way the composite is stressed (Barbosa and Kenny, 1999). The matrix must avoid the fibers from buckling in case of compressing loading, and therefore provides a very critical part of the composite. The reinforcing fibers could not bear any load when this fact is neglected. On the other side, high tensile loads without a matrix could be sustained by a bundle of continuous fibers in the direction of filaments. A transfer medium is also provided by the matrix; therefore, it does not give up its load-carrying capacity even if only one fiber breaks (Singh and Kamal, 1989).

6.4.2. Influence of Volume Fraction and Fiber Orientation

There is an important role of the volume fraction and fiber orientation on the mechanical and also other properties of the fiber-reinforced polymer composites. Two extreme situations are possible with respect to fiber orientation: totally random orientation and parallel alignment of the longitudinal axis of the fibers in a single direction. Figure 6.4(a) and (b) illustrate these two extremes, respectively. Generally, discontinuous fibers are randomly oriented and continuous ones are aligned. Inherent anisotropic thermal, dynamic, and mechanical properties are exhibited by the longitudinally aligned fiber-reinforced composites, where the reinforcements and maximum strength are obtained from the direction of fiber orientation. On the other side, transversely oriented fiber reinforcements happen to be virtually nonexistent, usually, fracture takes place at very low tensile stress, which might be even low than the pure matrix. Strength remains within these two extremes in randomly oriented fibrous composites. Thus, in the fiber-reinforced composites, the prediction of fiber orientation is much crucial in order to deduce the characteristics of the same.

6.4.2.1. Fiber Orientation in Injection Molded Fiber-Reinforced Composites

The fibers are assumed to be suspended in the matrix in injection molding and by that orient themselves in response to the mold cavity, kinematics of the flow, and other neighboring fibers.

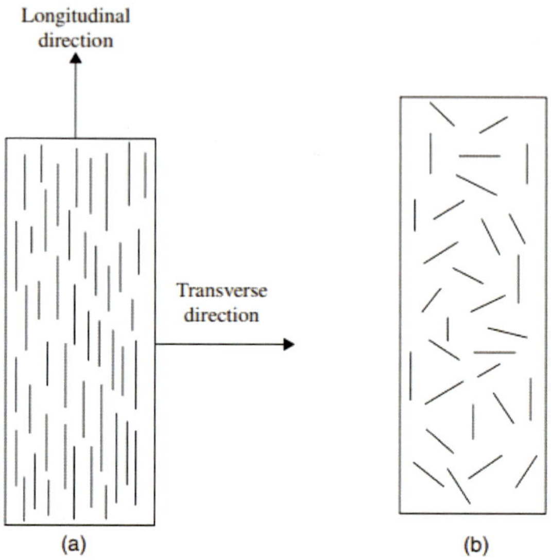

Figure 6.4. Diagrammatic representation of: (a) aligned and continuous; and (b) randomly oriented and discontinuous fiber-reinforced composites.

Source: https://onlinelibrary.wiley.com/doi/abs/10.1002/9783527645213.ch7.

However, an anisotropic behavior is often shown by the fiber suspensions because of a flow-induced fiber alignment in the direction of flow. Still, the characteristics of short fiber-reinforced thermoplastic composites undergo a problem connected to the fiber orientation, which following depends on the geometrical shape of the mold, e.g., inserts, gating, and section thickness and the processing conditions (Larsen, 2000). Normally, three flow regions exist throughout the filling of an injection molding die.

These regions are as follows (Malzahn and Schultz, 1986):

- A three-dimensional region closes to the gate;
- A fountain flow region upon the flow front;
- A lubrication region where most of the flow is accommodated and where no significant velocities from the main flow plane reside.

On fiber orientation, there is a complicated effect of flow behavior. But the two rules of thumb are determined (Gupta and Wang, 1993):

- Shearing flows favor to align fibers to the direction of flow;
- Stretching flows favor to align fibers to the direction of stretching. The stretching axis and the radial flow direction are perpendicular

to each other for a center gated disk, which can be seen in Figure 6.5.

In an injection molded parts, there are three different layers of the orientation of fibers that are as follows:

- During mold filling, a core is produced by in-plane fiber motion;
- Skin layers at the mold surface;
- Shell layers on one side of the core, along with a flow aligned orientation generated by gap-wise shearing.

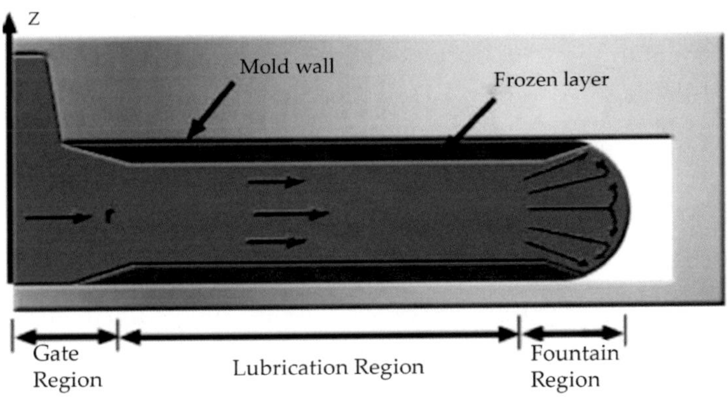

Figure 6.5. Flow regions along the injection molding process.

Source: https://www.meddeviceonline.com/doc/the-most-misunderstood-problem-in-medical-injection-molding-and-how-to-solve-it-0001.

The thermal and mechanical characteristics of the molded fiber-reinforced composite parts are influenced by this difference in fiber orientation in the different regions (Bailey, 1994; Chung and Kwon, 2002a, b).

Thus, it is of worth interest to envision the fiber orientation upon various processing conditions. To figure out the fiber orientation in injection-molded short fiber-reinforced composites under various molding conditions, there are several available reports that describe computer-aided simulation as well as analytical/numerical approach. In the planar flows, researchers recorded the nature of fiber orientation probability distribution function. In their study, they have stated the influence of fiber orientation upon the planar flow behavior regarding the short fiber-reinforced composites. Researchers also recorded the weld line strength of injection moldings produced through the co-injection molding method and the polymer melt flow. A comprehensive study is also proclaimed on the effect of processing variables ahead of the microstructure of injection molded short fiber-reinforced polypropylene (PP) composites (Wan et al., 2005; Senthilvelan and Gnanamoorthy, 2006).

6.4.3. Volume Fraction

In the envisioning of the characteristics of the fiber-reinforced composites, particularly hybrid composites, fiber volume fraction plays a vital role. Fiber volume fraction is expressed by the following equation.

V_f ¼ = V_F/V_C for hybrid composites

$V_f = V_f/V_c$ for normal composites

where; V_c refers to the total volume of composites; V_f signifies the volume fraction of fiber; and $V_F = V_{f1} + V_{f2}$. V_F represents the total volume fraction of both kinds of fibers. The matrix volume fraction is expressed by the equation given below:

$V_m = V_m/V_c$

where; V_m signifies the volume fraction of the matrix.

For a fiber-reinforced composite, to obtain the at most properties, an optimum spacing should be maintained, which is the at least acceptable spacing within the fibers, if it goes below the structure will going to disintegrate under loading since the tensile failure. This minimum spacing is described as the highest volume fraction acceptable for a composite.

6.4.4. Influence of Fiber Length

The strength of fiber-reinforced composites relies on the tensile strength and also depends on the scope to which applied stress is communicated to the fibers from the matrix. The scope of this load transmittance is a role of the strength of fiber-matrix interfacial bonds and fiber length. This fiber-matrix bond stops at the fiber ends under applied stress occurring a matrix deformation. At the fiber extremity, there is no stress transfer out of the matrix. Thus, the critical aspect ratio is being increased by a critical fiber length, length to diameter ratio (*L*/*D*) must be controlled in order to effective stress transfer within the matrix and the fiber at the interface (Nguyen-Chung et al., 1998; Nguyen-Chung and Mennig, 2001).

6.4.5. Influence of Voids

Air or other volatiles are captured in the material during the formation of laminates or during the inclusion of fibers into the matrix. The volatiles or trapped air presents in the cured laminate being a microvoid, which may influence significantly some of its mechanical characteristics. Some researchers investigate the origin of voids and discussed the different types of voids that come across in the composite and the ways to minimize the void content (Ishida and Koenig, 1979; Mallick, 1993). The most popular purpose of voids is the incapability of the matrix to evict all the air that is commenced in the yarn or roving when it infiltrates the resin impregnator. The eviction of the entrapped air will be affected by the viscosity of the resin, the rate at which the reinforcements infiltrate the matrix, the mechanical operation of the fibers, and the contact angle or wettability within the fiber surface and the matrix, etc. Usually, greater susceptibility leads to water diffusion, a high void content (beyond 20%) to lower fatigue resistance, and raised variation (scatter) results in mechanical properties (Patcharaphun and Mennig, 2005; Imihezri et al., 2006).

6.4.6. Influence of Coupling Agents

Same optimum physical properties are offered by the incorporation of the coupling agents and the fiber-reinforced composites. These properties of the composites are retained after environmental exposure. Changing the interface with different coupling agents can achieve good bonding at the interface. A significant approach for making better dispersibility and compatibility between the matrix and the filler is to make a hydrophobic coating of a compatible polymer upon the surface of the filler before weaving with the polymer matrix. Typically, coupling agents give better adhesion within the matrix and the filler. The choice of coupling agents that can offer both toughness and strength to a considerable degree is essential for composite material (Scola and Brooks, 1970; Belgacem et al., 1994).

6.5. SURFACE MODIFICATION

Currently, the key area of research to obtain optimum matrix/fiber properties is the surface modification of both fiber/ or matrix.

6.5.1. Surface Modification of Fibers

Reinforcing fibers can be altered by chemical and physical methods:

1. **Chemical Methods of Modification:** For the modification of fiber surface, more convenient ways are offered by the chemical methods. Chemical methods achieve compatibility within the

two polymeric materials, either presenting a third material into the composites that possess properties within those other two or using chemical reactions, creating some functional groups onto the fiber surface (Wakida and Tokino, 1996; Kaplan and Hansen, 1997).

2. **Physical Methods of Modification:** The surface and structural properties of the fibers are modified by the physical treatments, which in turn affect the mechanical bonding in the matrix-fiber interface. These physical treatments incorporate the electric discharge method through cold plasma or corona. In the situation of cold plasma treatment, a range of surface alterations such as cross-link could be proposed based on the nature and type of the gas utilized. In this technique, reactive free radicals upon the fiber surface could be formed, and the surface free energy could be decreased or increased. For Kevlar fiber, where there are deteriorating effects due to the moisture absorption, the plasma method is genetically an effective drying process giving additional benefits. One of the most compelling approaches for surface oxidation activation is corona treatment. The surface free energy of fibers is altered by this method (Park and Park, 1999; Lin et al., 2000).

It is reported by some scientists that the chemical treatment of Kevlar surface using Phosphoric acid significantly influence the degree of adhesion on interfaces between epoxy resin matrix and Kevlar fibers. While the interfacial shear strength (IFSS) of the Kevlar epoxy/fiber composites is enhanced by some scientists using chemical treatment of the fiber along with organic solvents (Yue and Padmanabhan, 1999; Park et al., 2002). Acid treatments being the modification tool for Kevlar, carbon fiber, and PBO are also proposed. A remarkable change is observed in the surface free energy of those altered fibers. The infrared mapping of the epoxy reacted epoxy/Kevlar system is also demonstrated. For the improvement of the fiber surface functionalities, various types of coupling agents are utilized to change the fiber surface. The most often used coupling agent is the silanes. Silanes offer dual functionalities in a single molecule; therefore, a bond is formed by the one part of the molecule to the fiber surface whereas the other part makes a bridge together with the matrix molecule (Figure 6.6) (Mavrich et al., 1995; Wu et al., 2004).

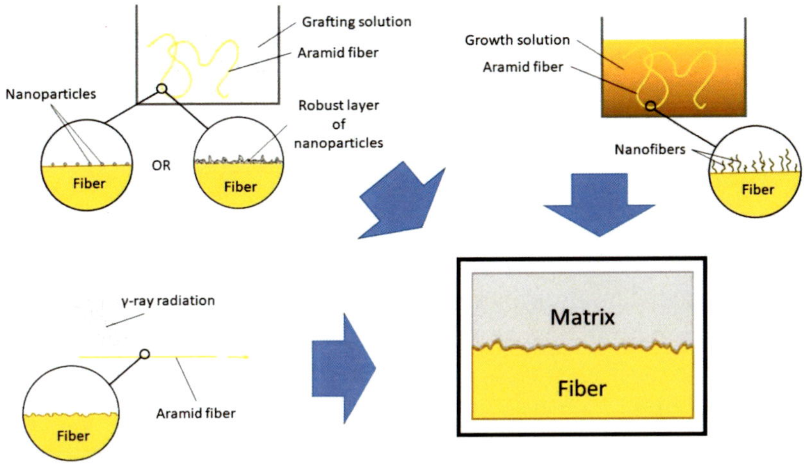

Figure 6.6. Surface modification of aramid fibers.

Source: https://www.mdpi.com/2079-6412/10/6/556/htm.

6.5.2. Surface Modification of Matrix Polymers

On the fiber surface, the wettability of the matrix polymer is based on the surface tension of both materials and the viscosity of the matrix. The surface tension of the matrix must be as at least as possible for the improved wetting of the matrix upon the fiber surface; it must be at least less as compared to the reinforcing fiber materials. Thus, following are the methods for the modification of the matrix within a fiber-reinforced composite (Kourkoulis and Andrianopoulos, 2000):

1. **Physical Methods:** It involves corona discharge treatment, plasma treatment, and UV irradiation.
2. **Chemical Treatments:** With the introduction of the polar groups to the matrix surface, the modification of the polymer surface can be possible. The surface properties are improved, and the polar groups are formed on the polymer surface if a polymer is dealt with the highly oxidative chemicals, for example, chromic acid/acetic acid, chromic anhydride/tetrachloroethane, chromic acid/sulfuric acid with suitable conditions (Farquhar, 1989).

6.5.3. Oxyfluorination and Fluorination as Polymer Surface Modification Tool

For polymeric materials, an exceptional role is played by the direct oxyfluorination and fluorination being a surface modification technique because it has various benefits that are as follows (Ai et al., 2007; Jeong et al., 2012):

- For the modification of the polymeric materials, elemental gas is used. The fluorination takes place at room temperature or low temperature because of the very high energy release at the time of the main elemental stages (Briggs et al., 1976). Any catalyst or initiation is not required.
- Gaseous mixtures of oxygen, fluorine, helium, nitrogen, chlorine, and chlorine, monofluoride might be utilized for direct fluorination.
- The reaction rewards by the diffusion-limited process.
- Over the range of ~0.01–10 *m*m, the thickness of the modified layer polymer can be forced to under control.
- The fluorine-treated plastics include a totally fluorinated, unprocessed virgin layer divided by a thin transition layer.
- Article of any shape can be changed.
- This technology is called dry one (just gases are used) and polymeric.
- One of the main benefits of direct oxyfluorination and fluorination is that just the thin surface layer of polymer is altered and therefore, the bulk properties of the polymer are not modified practically.

Oxyfluorination and fluorination serve with better properties as compared to the unmodified polymeric materials.

REFERENCES

1. Ai, T., Wang, R., & Zhou, W., (2007). Effect of grafting alkoxysilane on the surface properties of Kevlar fiber. *Polymer Composites, 28*(3), 412–416.
2. Akay, M., & Barkley, D., (1991). Fiber orientation and mechanical behavior in reinforced thermoplastic injection moldings. *Journal of Materials Science, 26*(10), 2731–2742.
3. Bailey, R. S., (1994). Conventional thermoplastics. *Handbook of Polymer-Fiber Composites, Polymer Science and Technology Series, 1*, 1–23.
4. Banks, R. E., (1994). In: Banks, R. E., Smart, B. E., & Tatlow, J. C., (eds.), *Organofluorine Chemistry* (Vol. 1, pp. 1–32).
5. Barbosa, S. E., & Kenny, J. M., (1999). Analysis of the relationship between processing conditions-fiber orientation-final properties in short fiber reinforced polypropylene. *Journal of Reinforced Plastics and Composites, 18*(5), 413–420.
6. Belgacem, M. N., Bataille, P., & Sapieha, S., (1994). Effect of corona modification on the mechanical properties of polypropylene/cellulose composites. *Journal of Applied Polymer Science, 53*(4), 379–385.
7. Blicblau, A. S., (2005). Transiting from school science to real engineering by employing materials science and engineering concepts within the school program. In: *4th ASEE/AaeE Global Colloquium on Engineering Education* (Vol. 4, pp. 1165). Australasian Association of Engineering Education.
8. Briggs, D., Brewis, D. M., & Konieczo, M. B., (1976). X-ray photoelectron spectroscopy studies of polymer surfaces. *Journal of Materials Science, 11*(7), 1270–1277.
9. Chou, T. W., (2005). *Microstructural Design of Fiber Composites* (Vol. 1, pp. 1–23). Cambridge University Press.
10. Chung, D. H., & Kwon, T. H., (2002a). Numerical studies of fiber suspensions in an axisymmetric radial diverging flow: The effects of modeling and numerical assumptions. *Journal of Non-Newtonian Fluid Mechanics, 107*(1–3), 67–96.
11. Chung, D. H., & Kwon, T. H., (2002b). Fiber orientation in the processing of polymer composites. *Korea-Australia Rheology Journal, 14*(4), 175–188.

12. Clegg, D. W., & Collyer, A. A., (1986). *Mechanical Properties of Reinforced Thermoplastics* (Vol. 326, pp. 1–33). London, UK: Elsevier Applied Science Publishers.
13. Farquhar, G., (1989). The defense research information center: Services and new developments. In: *Aslib Proceedings* (Vol. 1, pp. 1–23). MCB UP Ltd.
14. Fu, S. Y., Lauke, B., Mäder, E., Hu, X., & Yue, C. Y., (1999). Fracture resistance of short-glass-fiber-reinforced and short-carbon-fiber-reinforced polypropylene under Charpy impact load and its dependence on processing. *Journal of Materials Processing Technology, 89*, 501–507.
15. Gupta, M., & Wang, K. K., (1993). Fiber orientation and mechanical properties of short-fiber-reinforced injection-molded composites: Simulated and experimental results. *Polymer Composites, 14*(5), 367–382.
16. Herrera-Franco, P., & Valadez-Gonzalez, A., (2005). A study of the mechanical properties of short natural-fiber reinforced composites. *Composites Part B: Engineering, 36*(8), 597–608.
17. Imihezri, S. S. S., Sapuan, S. M., Sulaiman, S., Hamdan, M. M., Zainuddin, E. S., Osman, M. R., & Rahman, M. Z. A., (2006). Mould flow and component design analysis of polymeric based composite automotive clutch pedals. *Journal of Materials Processing Technology, 171*(3), 358–365.
18. Ishida, H., & Koenig, J. L., (1979). An investigation of the coupling agent/matrix interface of fiberglass reinforced plastics by Fourier transform infrared spectroscopy. *Journal of Polymer Science: Polymer Physics Edition, 17*(4), 615–626.
19. Jagur-Grodzinski, J., (1992). Modification of polymers under heterogeneous conditions. *Progress in Polymer Science, 17*(3), 361–415.
20. Jeong, E., Lee, B. H., Doh, S. J., Park, I. J., & Lee, Y. S., (2012). Multifunctional surface modification of an aramid fabric via direct fluorination. *Journal of Fluorine Chemistry, 141*, 69–75.
21. Joseph, K., Varghese, S., Kalaprasad, G., Thomas, S., Prasannakumari, L., Koshy, P., & Pavithran, C., (1996). Influence of interfacial adhesion on the mechanical properties and fracture behavior of short sisal fiber-reinforced polymer composites. *European Polymer Journal, 32*(10), 1243–1250.

22. Kadolph, S. J., (2002). In: Anna, & Langford, L., (eds.), *Textiles* (Vol. 1, pp. 1–22). Iowa State University.
23. Kaplan, S. L., & Hansen, W. P., (1997). Gas plasma treatment of Kevlar [R] and spectra [R] fabrics for advanced composites. In: *International Sampe Technical Conference* (Vol. 29, pp. 142–145). Sampe society for the advancement of material.
24. Kourkoulis, N. P., & Andrianopoulos, S., (2000). Plastically induced anisotropy on metal matrix composites. *Mechanics of Composite Materials and Structures, 7*(1), 1–17.
25. Larsen, Å., (2000). Injection molding of short fiber reinforced thermoplastics in a center-gated mold. *Polymer Composites, 21*(1), 51–64.
26. Lin, T. K., Wu, S. J., Lai, J. G., & Shyu, S. S., (2000). The effect of chemical treatment on reinforcement/matrix interaction in Kevlar-fiber/bismaleimide composites. *Composites Science and Technology, 60*(9), 1873–1878.
27. Mallick, P., (1993). *Fiber Reinforced Composites*. Marcel Dekker. New York.
28. Malzahn, J. C., & Schultz, J. M., (1986). Transverse core fiber alignment in short-fiber injection-molding. *Composites Science and Technology, 25*(3), 187–192.
29. Margrave, J. L., & Lagow, R. J., (1979). Direct fluorination: A 'new' approach to fluorine chemistry. *Prog. Inorg. Chem, 26*, 161.
30. Mavrich, A. M., Ishida, H., & Koenig, J. L., (1995). Infrared mapping of surface-modified Kevlar fiber-reinforced epoxy systems. *Applied Spectroscopy, 49*(2), 149–155.
31. Miraftab, M., Qiao, Q., Kennedy, J. F., Anand, S. C., & Groocock, M. R., (2003). Fibers for wound dressings based on mixed carbohydrate polymer fibers. *Carbohydrate Polymers, 53*(3), 225–231.
32. Modzelewski, T., (2015). *Design and Synthesis of Polyphosphazenes: Hard Tissue Scaffolding Biomaterials and Physically Crosslinked Elastomers, 1*, 1–23.
33. Mohanty, S., Nayak, S. K., Verma, S. K., & Tripathy, S. S., (2004). Effect of MAPP as a coupling agent on the performance of jute-PP composites. *Journal of Reinforced Plastics and Composites, 23*(6), 625–637.

34. Morley, J. G., (1976). New types of reinforced composite materials. *Physics Reports, 28*(3), 245–302.
35. Mukherjee, M., (2007). *Effect of Fluorination and Oxy-Fluorination of Kevlar Fiber* (Vol. 1, pp. 1–28.) (Doctoral dissertation, IIT, Kharagpur).
36. Nguyen-Chung, T., & Mennig, G., (2001). Non-isothermal transient flow and molecular orientation during injection mold filling. *Rheologica Acta, 40*(1), 67–73.
37. Nguyen-Chung, T., Plichta, C., & Mennig, G., (1998). Flow disturbance in polymer melt behind an obstacle. *Rheologica Acta, 37*(3), 299–305.
38. Park, S. J., & Park, B. J., (1999). Electrochemically modified PAN carbon fibers and interfacial adhesion in epoxy-resin composites. *Journal of Materials Science Letters, 18*(1), 47–49.
39. Park, S. J., Seo, M. K., Ma, T. J., & Lee, D. R., (2002). Effect of chemical treatment of Kevlar fibers on mechanical interfacial properties of composites. *Journal of Colloid and Interface Science, 252*(1), 249–255.
40. Patcharaphun, S., & Mennig, G., (2005). Properties enhancement of short glass fiber-reinforced thermoplastics via sandwich injection molding. *Polymer Composites, 26*(6), 823–831.
41. Peters, S. T., (2013). *Handbook of Composites* (Vol. 1, pp. 1–23). Springer Science & Business Media.
42. Sadov, F. I., Korchagin, M. V., & Matetskii, A. I., (1978). *Chemical Technology of Fibrous Materials, 1*, 1–20.
43. Scola, D. A., & Brooks, C. S., (1970). Surface aspects of new fibers, boron, silicon carbide, and graphite. *The Journal of Adhesion, 2*(3), 213–237.
44. Senthilvelan, S., & Gnanamoorthy, R., (2006). Fiber reinforcement in injection-molded nylon 6/6 spur gears. *Applied Composite Materials, 13*(4), 237–248.
45. Singh, P., & Kamal, M. R., (1989). The effect of processing variables on microstructure of injection molded short fiber reinforced polypropylene composites. *Polymer Composites, 10*(5), 344–351.
46. Wakida, T., & Tokino, S., (1996). *Surface Modification of Fiber and Polymeric Materials by Discharge Treatment and its Application to Textile Processing, 1*, 1–33.

47. Wan, Z. H., Lin, J. Z., & You, Z. J., (2005). non-axisymmetric instability in the Taylor-Couette flow of fiber suspension. *Journal of Zhejiang University-Science A, 6*(1), 1–7.

48. Wu, G. M., Hung, C. H., You, J. H., & Liu, S. J., (2004). Surface modification of reinforcement fibers for composites by acid treatments. *Journal of Polymer Research, 11*(1), 31–36.

49. Yue, C. Y., & Padmanabhan, K., (1999). Interfacial studies on surface-modified Kevlar fiber/epoxy matrix composites. *Composites Part B: Engineering, 30*(2), 205–217.

50. Zhang, H., Farris, R. J., & Westmoreland, P. R., (2003). Low flammability and thermal decomposition behavior of poly(3,3'-dihydroxybiphenyl isophthalamide) and its derivatives. *Macromolecules, 36*(11), 3944–3954.

Chapter

7

Nylon Fiber-Reinforced Polymeric Materials

CONTENTS

7.1. Introduction..218

7.2. Nylon Fibers Used as Reinforcements...219

7.3. Matrices and Applications..223

7.4. Manufacturing af Nylon-Reinforced Composites..............................227

References...232

7.1. INTRODUCTION

Generally, composites may be defined as the consolidation of two or more materials coming out of from the combination of some fundamental structural material and a second substance, i.e., the matrix [1, 2]. There is the capacity of applying the desirable characteristics of each component based on the concept of composite, getting a material with an increased performance according to the isolated constituents (Bailey et al., 1998; Kroschwitz and Seidel, 2007).

Various morphologies can be appeared for the incorporated material, for instance, fibers, particles, lamellae, or whiskers, but they always have the same job, which is to give their own beneficial mechanical properties to the matrix material. The key supplement variables are the morphology and the chemical nature of the reinforcing agent that offer a competitive profit to composites upon homogeneous materials. In fact, the opportunity is given by the optimization and a wise choice of these features, to deal with physically uncorrelated parameters like density, strength, cost, and electrical properties. Within the available shapes of the reinforcement particles, specifically, fibers are attractive for their very anisotropic nature and high aspect ratio (Robinson and Robinson, 1994). If needed, this allows conveying anisotropic physical properties to the complete composite. This chapter will discuss only fiber-reinforced composites.

Three examples may be confronted about the relative parts of the components in a composite material (Kim and Shim, 2009):

- Firstly, the reinforcement possesses high stiffness and strength, on the other side, the matrix must pass on the stress from one reinforcement fiber to another. This is the instance of high-performance composites, where high strength reinforcement fibers are exploited in calibrating their orientation, high volume fractions, and dispersal for optimum physical mechanical property development.
- In some other situations, the matrix has enviable intrinsic physical, processing, and chemical properties already, and inclusion of the reinforcement is engaged to make better other valuable technological properties like a creep, tear resistance, or tensile strength. The need of developing the engineering properties of a matrix drives the formation of this category of materials in order to broaden its scope of applications. Usually, the moderate

concentration of fibers being flake or discontinuous random fibers is applied for these composites.

- Thirdly, the matrix might be a high-performance material, although with low process ability, high cost, or not enough esthetical appeal. In this example, the objective of the inclusion of the fibers is to maintain adequate performance, while resolving the weakness of the matrix material, particularly under a cost-effectiveness view.

The properties desired for a good reinforcing filler are that: it is strong, stiff, and light, and it likely changes the failure process in a beneficial way. The stiffness, strength, and specific gravity of many different reinforcement fibers are compared in Table 7.1.

The performance of the nylon is good between polymeric reinforcement fibers, and albeit its performance is substandard to that of carbon and inorganic fibers, but it provides a really attractive cost-advantage balance.

Table 7.1. Representative Properties of Various Reinforcement Fibers

Material	Strength (GPa)	Modulus (GPa)	Typical Diameter (mm)	Specific Gravity
Alumina	1–3	400–500	3–20	3.34
Aramid	3	100–150	12–15	1.44
E-glass	3	70	10	2.6
Nylon	1	6	3–500	1.1
PAN-based carbon	2–3	200–500	7–10	1.7–2
Polyolefin	2	1–5	5–500	0.97
Rayon	1	7	3–500	1.5
S-glass	5	90	10	2.5
Silicon carbide	2–6	300–400	10–100	3.2

7.2. NYLON FIBERS USED AS REINFORCEMENTS

The general class of nylon consists of all linear polyamide (PA) polymers obtained from aliphatic monomers. There are further two groups of linear aliphatic PAs: those created from the poly-condensation reaction of dicarboxylic and diamines acids, and those synthesized using amino carboxylic acids (either lactams or v-amino carboxylic acids, through ring-opening polymerization). For the indication of the number of carbon atoms within the monomeric building blocks, the numbers are used by the common nomenclature of nylon. For example, nylon 6 refers to the polymer of caprolactone [H_2N $(CH_2)_5COOH$]. Nylon 10,6 signifies the sebacic acid [$HOOC(CH_2)_8COOH$], and polycondensate of hexamethylenediamine [H_2N $(CH_2)_6NH_2$].

The first nylons are formed in the 1930s and commercialized from 1939, by taking specific advantage of the demand for raw materials for the purpose of military use in the time period of the Second World War. Since then, a very versatile material has been provided by the nylon fibers for enormous applications, due to their remarkable properties, particularly because of the voluminous hydrogen bonding within the polymer chains. The reasons for the attractiveness of the nylon fibers are their abrasion resistance, durability, and high elongation and tenacity. Through nylon fibers, the highly polar nature of the amide groups supports the absorption of moisture, which, when in contact with water, therefore, endure poor dimensional stability. While processing the nylon, strict attention should be given to avoid its exposure to the water. Purity is also a problem as reactive impurities could initiate uncontrolled chain growth, minimizing the quality of the product, and turning yellow. At high temperatures, nylon is sensitive to oxidation, hence, oxygen should be evacuated during fiber spinning and polymerization. Additives can be included to control the sensitivity of nylon to specific conditions and their activity. Light protection and aging can be provided by titanium dioxide or manganese (II) compounds. Heat aging resistance for applications like tire cords is obtained with antioxidants, e.g., substituted aromatic diamines (Lu et al., 2006).

A PA appropriate for spinning mainly encompasses methylene groups within the amide groups. In an aliphatic PA chain, cyclic segments increase their melting point and stiffen them. An increase in the solubility in organic solvents and a minimization of the melting temperature are brought up by the introduction of the side carbon-carbon chains in the polymer. In low molecular mass nylon, with the increase of the average molecular weight, the melting point grows. Though, in the situation of high molecular weight polymers, this effect fades and can be neglected. A raise in the molecular mass also causes a growth in the strength of the fibers. An extreme limit that can be applied, in the molecular mass, is posed along with the viscosity of the melt, which grows by increasing molecular weight, without permitting melt spinning. The quantity of methylene groups in the chain is another factor that affects the mechanical, processing, and physical properties of nylon materials. With the increase of the ratio of methylene to carbonamide groups, there is a decrease in the melting point of nylon (Francis et al., 2010).

Nylon 12, nylon 11, nylon 7, and nylon 6 are the nylons of the amino carboxylic type appropriate for fiber production. Nylon 12 and nylon 11, albeit appropriate for fiber production, are pretty expensive, and their usage is limited to a few peculiar applications like transmission belts, luggage, or ropes. Because of the easy polymerization-spinning process, high yield of synthesis, low water absorption, and very good tensile properties, nylon 7 would be more adapted. Though, economical industrial manufacturing of nylon 7 is not allowed by the high cost of monomer. Nylon 6 is the reason for about half of the worldwide production of nylon fibers, as its caprolactam, monomer, is easily attainable by cheap petrochemicals. Though nylon 6 is not the perfect material for the production of the fiber due to the difficulty in the transformation of the monomer into a continuously spinnable polymeric melt, and due to its pretty high-water absorption (Iijima, 1991).

Undoubtedly, the most generally employed nylons derived from the polycondensation of dicarboxylic acids and diamines are nylon 6,10 and nylon 6,6. For fiber production, nylon 6,6 along with nylon 6 is the reason for around the total consumption of nylon, because of its competitive performance and attractive cost. Nylon 6,10 would be specifically suited for usage in fiber preparation because of its high elasticity and low water absorption, but its higher cost according to nylon 6,6 avoided broad commercial exploitation.

At temperatures above around 300°C, all nylon fibers of commercial concern are melt-spun, by spinnerets having holes 200–400 mm in diameter. The process can be detached into various subsequent phases, or it can be executed constantly from the monomer to the spun fiber. After that, filaments are linked into yarns. For the determination of the structure obtained by the PA molecules, clearly processing parameters, e.g., spinning temperature or draw ratio are essential, and thus, the mechanical and physical properties (Pang and Fancey, 2009). The ultimate diameter of the fibers is determined by the draw ratio and spinneret dimensions. For the purpose of reinforcement, about 1–20 dtex filaments are appropriate. Somewhat coarse denier yarns are required by the high strength applications, e.g., tire-cords, ranging from 400 to above 35000 dtex. A nylon fiber being a reinforcement has a typical diameter ranges from 3 to 500 mm and relies much on the application (Mahfuz et al., 2006). There is also an importance of the length of the fibers. As compared to the production of continuous fibers-containing composites, adding short fibers within a polymeric matrix is cheaper and easier. Although, usually, short fibers produce materials having lower strength, fracture toughness, and stiffness than constant fibers (Shi et al., 2004).

Fibers can also be improved by covering them with a material that includes new characteristics to the entire fibrous material. Moreover, if designed correctly, the surface characteristics of the fibers can be influenced by coating, which is important for a good reinforcement, as an optimal interfacial adhesion within the filler and the matrix is essential for a proper transfer of stress from the previous to the latest. For an efficient pass on of the applied load, interfacial bonding should be adequately strong, but not too much, since it could also minimize the toughness of the composites and push for crack propagation across the fibers (Zhou et al., 2003). On the other hand, a poor interfacial adhesion is not enviable on the account of the weak interface that would facilitate being a preferential direction of crack propagation. In this situation, a deformable interface that permits fiber to withdraw and debonding can influence positively the toughness, albeit at the cost of the strength of the composite (Wang et al., 2010).

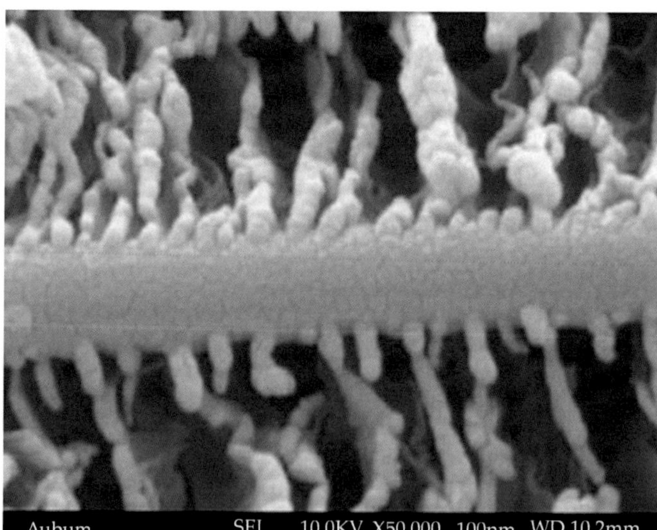

Figure 7.1. Rod-shaped Si_3N_4 particle involving a nylon 6 matrix.

Source: https://www.sciencedirect.com/science/article/abs/pii/S092150930801071X.

Researchers have been reported that conductivity is provided to a nylon 6,6 fabric by coating it with a silver nano-crystalline thin film (i.e., from 220 to 2800 nm) using radio-frequency sputtering technique (Rangari et al., 2009).

Nylon 6 polymer chains are shown in Figure 7.1 that are entrapped on the rod-shaped filler particle. Based on the reinforcement of nylon through aramid fibers, this effect is identical to the creation of a trans-crystalline layer (Feldman et al., 2006; Causin et al., 2009). The reinforcement influence of these particles is used to its generous extent when the matrix is capable of crystallization, or else coat with a layer, i.e., the filler particles (Figure 7.2) (Caskey et al., 2003; Causin et al., 2008).

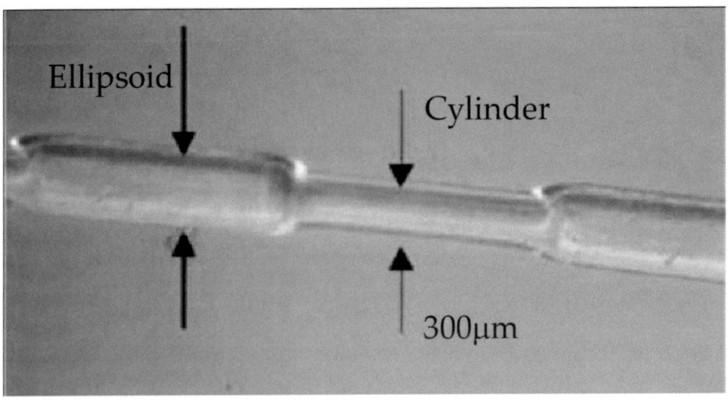

Figure 7.2. Optical micrograph of a chain-shaped fiber.

Source: https://www.sciencedirect.com/science/article/abs/pii/S1359646206000637.

For various embedded depths, single fiber pull-out tests have been performed, and such a relevant effect of the chain-shape geometry is shown on the characteristics of the composites (Ma and Chen, 1992; Goitisolo et al., 2010a). Multiple peaks are shown by the pulling-out load vs displacement curves of the chain-shaped fiber having composites that do not exist in the material holding normal straight short fibers (see Figure 7.3).

7.3. MATRICES AND APPLICATIONS

Table 7.1 expresses that the mechanical and physical properties of nylon are phenomenal but not exceptional, e.g., compared to carbon, inorganic, or aramid fibers. Hence, for the reinforcement of elastomers and thermoplastic polymers, nylon is the most suitable (Goitisolo et al., 2010b). The potential in property betterment of nylon fibers is shown by the recent literature reports particularly in polyethylene terephthalate (PET), polymethylmethacrylate (PMMA), rubber, and polycarbonate (PC) (Younan et al., 1992).

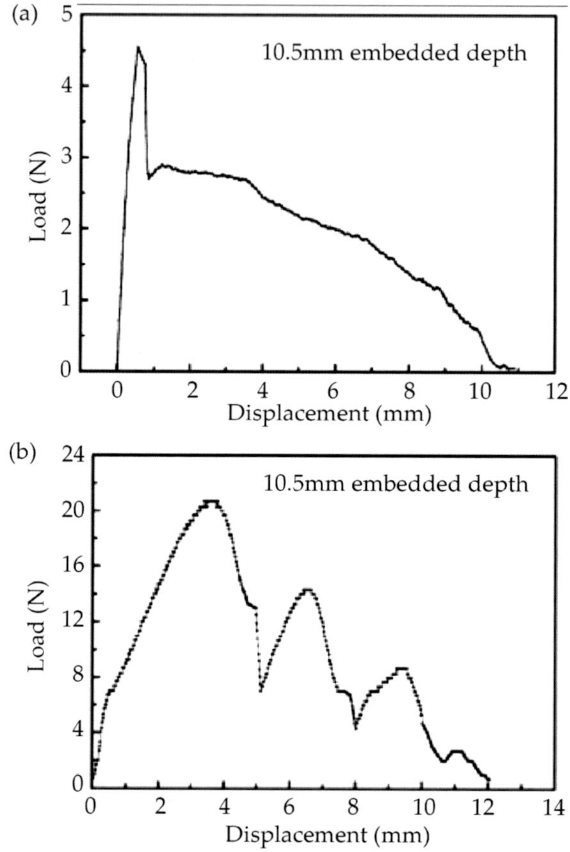

Figure 7.3. Pull-out load versus displacement curves: (a) straight fibers; and (b) chain-shaped fibers.

Source: https://onlinelibrary.wiley.com/doi/10.1002/9783527645213.ch9.

The work of one of the researchers can be cited among many related studies on the effect of short fibers on a rubber matrix (Saad and Younan, 1995). Acrylonitrile butadiene rubber (NBR)-based composites have been prepared by them by combining short nylon fibers of various lengths (10, 6, or 2 mm) and concentration into the matrix utilizing a two-roll mixing mill in accordance with a base formulation. The curing properties of the samples are studied, the effect of loading, fiber length, and rubber cross-linking systems are noticed on the characteristics of the composites. Good reinforcement is offered by the inclusion of the nylon fibers to NBR, bringing betterment in mechanical characteristics. It has been found that a fiber length of 6 mm is optimal for the perfect balance of properties. There is a maximum load transmittance between the matrix and the fiber at this critical fiber length. When the critical fiber length is more than the fiber length, then the stressed fiber is going to be de-bonded from the matrix, and the composite eventually fails at low load. When the critical fiber length is below fiber length, then the stressed composites result in the breaking of the fiber. Furthermore, the elongation at break is minimized within crease in the length of the fiber as long fibers degrade the probability and possibility for deformation and rearrangement of the matrix at applied load (Rajesh et al., 2004).

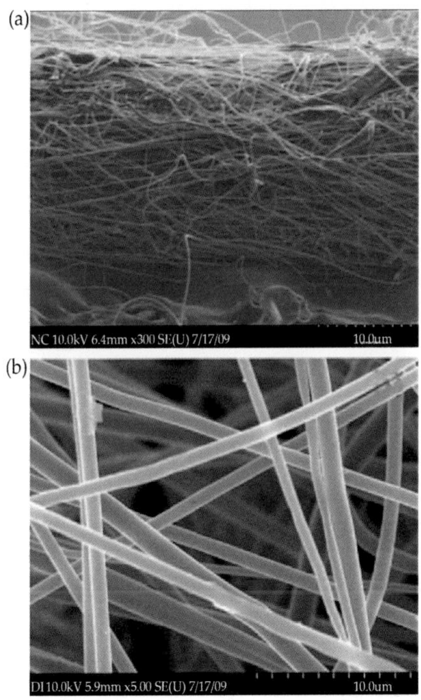

Figure 7.4. FESEM micrographs of: (a) the cross-section of a nylon 6 electrospun fiber mat; and (b) of a detail of the electrospun fibers.

Source: https://www.sciencedirect.com/science/article/abs/pii/S0014305710000340.

The quantity of the fibers should be selected accurately because their presence makes a rise in torque and viscosity of the mixture, if excessive, it can turn to brittleness. It has been observed by the scientists that when the fiber loading increases, the brittleness, and stiffness of the composite increases slowly, along with an associated fall in elongation at break (Seema and Kutty, 2006).

Surprisingly, the mechanical properties are modified by the existence of fibers in a rubber matrix. Moreover, they also affect the response of the material to curing. Hence, the design of a process containing fiber-reinforced rubber should also focus on the optimization of the curing step (Figure 7.4) (Naskara et al., 2004).

Note that there is an observation of the marked influence of the diameter of the fiber on the range of interfacial adhesion. The fibers having a larger diameter are shown in Figure 7.5(a), which are displaying a much feeble interfacial adhesion. The fibers that are oriented perpendicularly to the fracture surface are not be subjected to pulling out, while the fibers that are oriented tangentially to the fracture surface are pulled out as well as hanged loose, particularly on the formation of bundles. The quantity of very fine filaments also exists along with such large fibers. These come up by embedded strongly in the matrix (refers to Figure 7.5(b) and (c)), the bare ends of the fibers stick out directly from inside the matrix, without the existence of craters at their base because of debonding (Martınez-Barrera et al., 2008, 2010).

(a)

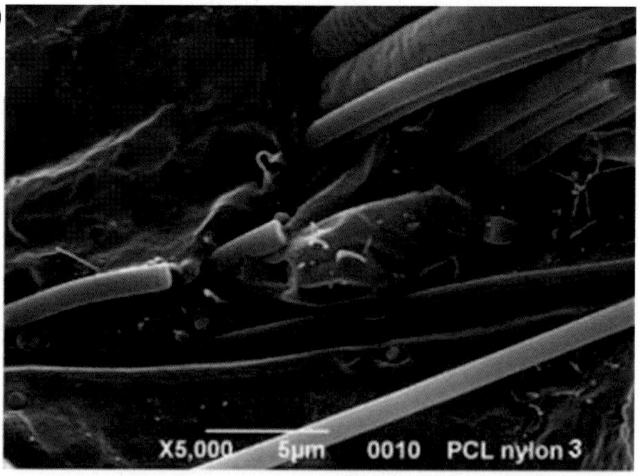

(b)

(c)

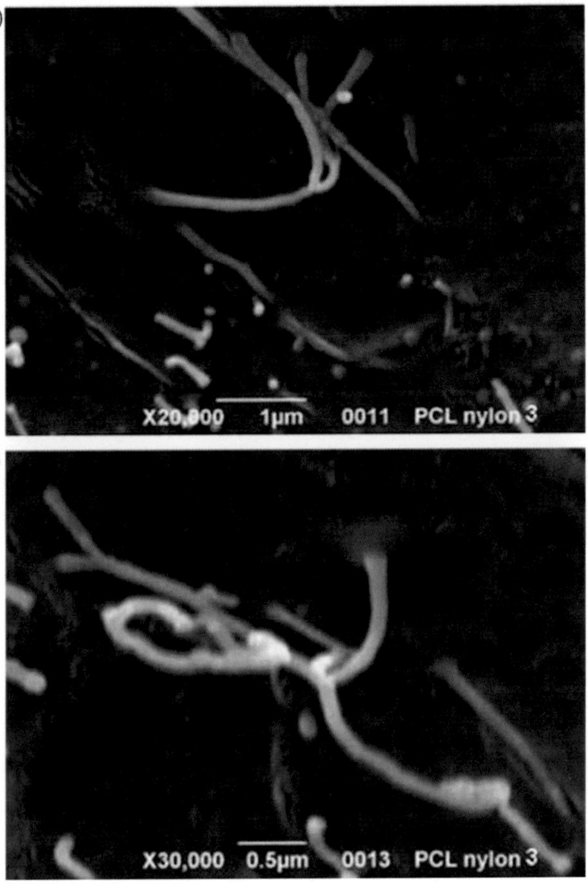

Figure 7.5. SEM micrographs at various magnification of the cryogenic fracture surface of a sample of PCL involving nylon 6 electrospun fibers.

Source: https://www.researchgate.net/publication/239141322_Polye-caprolactone_filled_with_electrospun_nylon_fibers_A_model_for_a_facile_composite_fabrication.

7.4. MANUFACTURING OF NYLON-REINFORCED COMPOSITES

Despite of the particular technology, the process of creating nylon reinforced composites needs a first step, while in a fluid state, where the matrix is compelled to enclose the fibers, and the next step where under controlled conditions, the matrix is solidified to keep the dispersion of the fibers obtained in the previous phase.

For thermoset matrices, the straightforward approach involves percolating a fluid prepolymer inside a textile or fiber mat, pursued by curing using the application of heat to huddle the composite the required structure and shape. Being a starting material, "Prepreg" sheets are much generally used. Usually, they are unidirectional (UD) fabrics or tapes formed of fibers bonded with a thin film of partly cured resin. The same idea is involved when treating thermoplastic matrices, excluding that the fluid format of the matrix is found through melting. Pressure and processing temperature are in these examples crucial factors since they should be large enough to support a considerable viscosity to the liquid, without causing to happen degradative side reactions (Hejazi et al., 2008; Akangah et al., 2010).

The basic concepts listed above have grown into more refined approaches where a dry fiber preform is settled into a closed mold following the infusion of the precursor through applying pressure or vacuum, often below 10 bar. There is a demoldation of the composite product after cross-linking or polymerization. Pultrusion is another technique that is significant for its technological implications, where continuous filaments are impregnated and designed by a matrix bath and after that by a die. The die facilitates orient the reinforcement, calibrating the amount of the matrix enclosing the fiber, and to pass on a shape to the laminate. For a thermoset matrix, the cure can be finished in the die, or it can be accomplished at a second turn (Mark et al., 1994; De and White, 2001).

For the preparation of composite, the selection of the most capable processing method is very substantial for obtaining the required performances. For example, recently a method of sample preparation is proclaimed by some researchers in which tension is practiced to the reinforcing fibers, that is supposed to be released before the addition of the fibers within the matrix. The viscoelastically strained fibers are the reason for the impartation of the compressive stresses during the solidification of the matrix, which makes better the mechanical and physical properties. Manufacturing polyester/nylon and epoxy/nylon composites by this process, the flexural modulus is raised by 50%, corresponding to control samples prepared without prestressing the fibers. This example exhibits the intrinsic potential of processing where all the formulation conditions are equal, for achieving essential advancements in the final characteristics of the composite (Cheremisinoff, 1993). A micrograph of a PMMA sample is shown in Figure

7.6 which is reinforced using nylon fibers found by *in-situ* polymerization cooperated by supercritical CO_2. Kinetically trapped blends of immiscible polymers are allowed to be created by this approach that is homogeneous upon the nanometer length scale, therefore, the matrix entirely fills the gap within the reinforcing fibers (refers to Figure 7.6).

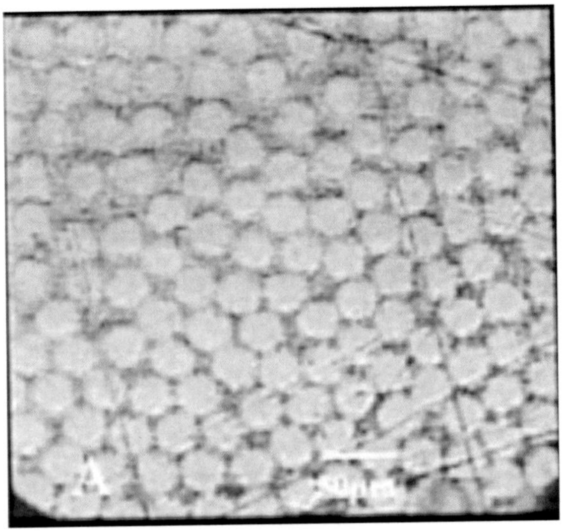

Figure 7.6. Optical micrograph of a PMMA/nylon composite formed by in-situ polymerization in supercritical CO_2

Source: https://onlinelibrary.wiley.com/doi/abs/10.1002/pc.10051.

In the impregnation step, the fluidity of the matrix material favors taking complete advantage of the remarkably efficient manufacturing methods that operate vacuum infusion of the mold or injection molding. UV radiation or heat can initiate polymerization and might need the inclusion of a catalyst system, which can be combined with the precursor before impregnation (Dogan et al., 2008).

When treating semicrystalline polymers, caution should be taken in selecting the processing temperature to balance the rates of crystallization (lessened at high temperature) and polymerization (preferred by increasing temperature). At a very low temperature, there will be a fast crystallization and the reactive chain expires, and monomers can be caught in a trap within the crystals prior to their polymerization. Whereas, on the very high temperature, there is a reduction of the final degree of crystallinity which minimizes the stiffness, strength, and chemical resistance of the polymer

(Fong, 2004). The potential is particularly attractive for forming transparent nanocomposites on the basis of PMMA using electrospinning, since the alternate techniques for reinforcing this polymer, e.g., the inclusion of rubber-particles or nanoparticles, the formulation of polymer alloys, or the conventional fiber reinforcement, normally turned out to have minimum transparent materials because of the mismatches of refractive index (Figure 7.7) (John et al., 2001).

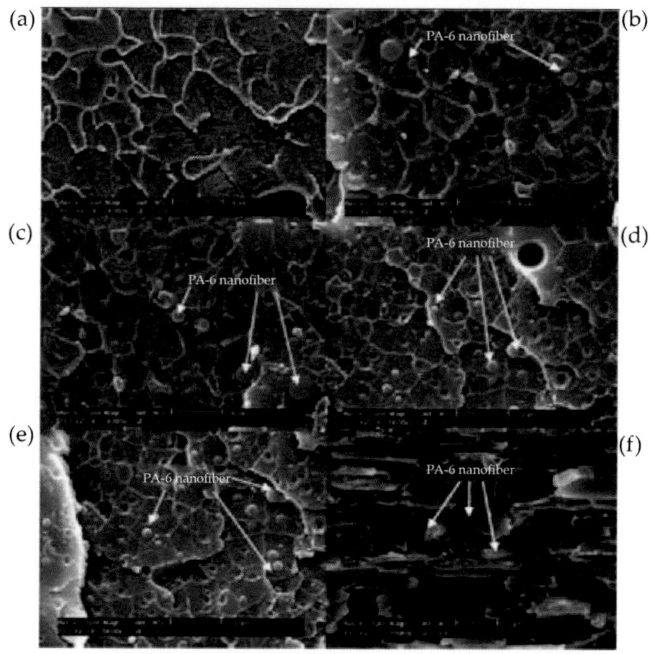

Figure 7.7. Fracture morphology of nylon 6/6 composite, and (f) PMMA/2.5% nylon 6 fiber-reinforced PMMA composites: (a) pure composite (tensile fracture surface) Reprinted PMMA; (b) PMMA/0.5% nylon 6 composite; (c) PMMA/1.5% nylon 6 composite; (d) PMMA/2.5% nylon 6 composite; (e) PMMA/3.5% nylon.

Source: https://onlinelibrary.wiley.com/doi/abs/10.1002/pc.20551.

A very interesting alternative approach to make fibrous reinforcements inside a polymer matrix is using an *in-situ* fibrillation method (Subbiah et al., 2005). Three fundamental steps that are needed to achieve microfibrils inside a polymer mixture are as follows (Tian et al., 2007):

- Melt blending with the expulsion of two immiscible polymers possessing different melting temperatures (i.e., mixing step);

- Cold drawing of the extrudate having a good orientation of both components (i.e., fibrillation step);
- Thermal treatment at a temperature within the melting points of both blend partners (i.e., isotropization step) occurring while processing the drawn blend through compression molding or injection.

A micrograph of the so-found nylon fibrils that dispersed in the PET matrix is shown in Figure 7.8. Furthermore, the researchers have been able to use the potential of nylon to exfoliate clay, and hence reinforce the material using the *in-situ* formation of nanofibrils and also by manufacturing clay-based nanocomposites at the same time. The increase that is observed in the elasticity modulus (modulus grows by 38% corresponding to the PET matrix, having just 1.96% MMT in the entire nanocomposite) is greater as compared to any previously noticed PET-based composites having nanoclay, because of the too good dispersion of the nanoclay and the synergistic effect of the nanofibrils (Huang et al., 2003; Neppalli et al., 2010).

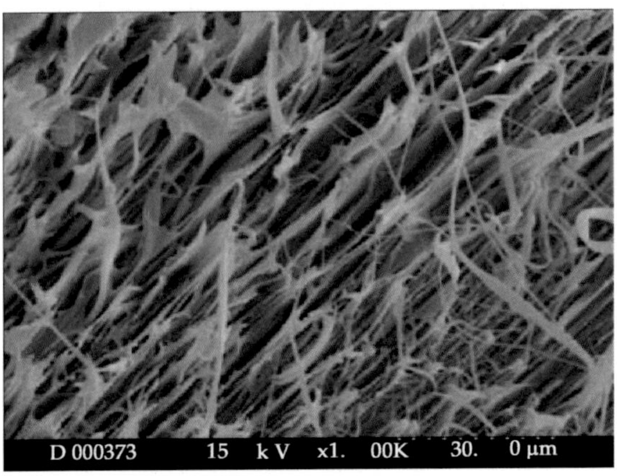

Figure 7.8. SEM micrograph of nylon fibrils achieved in-situ in a PET matrix.

Source: https://onlinelibrary.wiley.com/doi/abs/10.1002/mame.200900263.

REFERENCES

1. Akangah, P., Lingaiah, S., & Shivakumar, K., (2010). Effect of nylon-66 nano-fiber interleaving on impact damage resistance of epoxy/carbon fiber composite laminates. *Composite Structures, 92*, 1432–1439.
2. Bailey, J. E., Brinder, J., & Bohnet, M., (1998). *Ullmann's Encyclopedia of Industrial Chemistry* (6th edn., Vol. 1, pp. 1–21). Wiley-VCH, Weinheim.
3. Caskey, T. C., Lesser, A. J., & McCarthy, T. J., (2003). Evaluating the mechanical performance of supercritical CO_2 fabricated polyamide 6,6/PMMA, fiber-reinforced composites. *Polymer Composites, 24*, 545–554.
4. Causin, V., Yang, B. X., Marega, C., Goh, S. H., & Marigo, A., (2009). Nucleation, structure, and lamellar morphology of isotactic polypropylene filled with polypropylene-grafted multiwalled carbon nanotubes. *European Polymer Journal, 45*, 2155–2163.
5. Causin, V., Yang, B. X., Marega, C., Goh, S. H., & Marigo, A., (2008). Structure-property relationship in polyethylene reinforced by polyethylene grafted multiwalled carbon nanotubes. *Journal of Nanoscience and Nanotechnology, 8,* 1790–1796.
6. Cheremisinoff, N. P., (1993). *Elastomer Technology Handbook.* CRC Press, Boca Raton.
7. De, S. K., & White, J. R., (2001). *Rubber Technologist's Handbook.* Rapra Technology, Shrewsbury, UK.
8. Dogan, O. M., Bolayır, G., Keskin, S., Dogan, A., & Bulent, B., (2008). The€ evaluation of some flexural properties of a denture base resin reinforced with various aesthetic fibers. *Journal of Materials Science-Materials in Medicine, 19,* 2343–2349.
9. Feldman, A. Y., Wachtel, E., Zafeiropoulos, N. E., Schneider, K., Stamm, M., Davies, R. J., et al., (2006). *In situ* synchrotron microbeam analysis of the stiffness of trans crystallinity in aramid fiber reinforced nylon 66 composites. *Composites Science and Technology, 66*, 2009–2015.
10. Fong, H., (2004). Electrospun nylon 6 nanofiber reinforced BIS-GMA/TEGDMA dental restorative composite resins. *Polymer, 45*, 2427–2432.
11. Francis, L., Giunco, F., Balakrishnan, A., & Marsano, E., (2010). Synthesis, characterization, and mechanical properties of nylon-silver

composite nanofibers prepared by electrospinning. *Current Applied Physics, 10*, 1005–1008.

12. Goitisolo, I., Eguiazabal, J. I., & Nazabal, J., (2010a). Stiffening of poly(ethylene terephthalate). by means of polyamide 6 nanocomposite fibers produced during processing. *Composites Science and Technology, 70*, 873–878.
13. Goitisolo, I., Eguiazabal, J. I., & Nazabal, J., (2010b). Stiffening of PC by addition of a highly dispersed and fibrillated amorphous polyamide-based nanocomposite. *Macromolecular Materials and Engineering, 295*, 233–242.
14. Hejazi, S. M., Abtahi, S. M., Sheikhzadeh, M., & Semnani, D., (2008). Introducing two simple models for predicting fiber-reinforced asphalt concrete behavior during longitudinal loads. *Journal of Applied Polymer Science, 109*, 2872–2881.
15. Huang, Z. M., Zhang, Y. Z., Kotakic, M., & Ramakrishna, S., (2003). A review on polymer nanofibers by electrospinning and their applications in nanocomposites. *Composites Science and Technology, 63*, 2223–2253.
16. Iijima, S., (1991). Helical microtubules of graphitic carbon. *Nature, 354*, 56–58
17. John, J., Gangadhar, S. A., & Shah, I., (2001). Flexural strength of heat polymerized polymethylmethacrylate denture resin reinforced with glass, aramid, or nylon fibers. *The Journal of Prosthetic Dentistry, 86*, 424–427.
18. Kim, S. J., & Shim, S. B., (2009). The effect of the processing factors on the physical properties of high shrinkable nylon composite yarns. *Fiber Polymers, 10*, 813–821.
19. Kroschwitz, J. I., & Seidel, A., (2007). *Kirk-Othmer Encyclopedia of Chemical Technology* (Vol. 1, pp. 1–23). Wiley-Interscience, Hoboken, NJ.
20. Lu, X., Zhang, Y., & Xu, J., (2006). Influence of fiber morphology in pull-out process of chain-shaped fiber reinforced polymer composites. *Scripta Materialia, 54*, 1617–1621.
21. Ma, C. M., & Chen, C. H., (1992). Pultruded fiber-reinforced poly(methyl methacrylate). composites. II Static and dynamic mechanical properties, environmental effects, and post formability. *Journal of Applied Polymer Science, 44*, 819–827.

22. Mahfuz, H., Adnan, A., Rangari, V. K., Hasan, M. M., Jeelani, S., Wright, W. J., et al., (2006). Enhancement of strength and stiffness of nylon 6 filaments through carbon nanotubes reinforcement. *Applied Physics Letters, 88*, 083119.
23. Mark, J. B., Erman, B., & Eirich, F. R., (1994). *Science and Technology of Rubber*. Academic Press, New York.
24. Martınez-Barrera, G., Giraldo, L. F., Lopez, B. L., & Brostow, W., (2008). Effects of c radiation on fiber-reinforced polymer concrete. *Polymer Composites, 29*, 1244–1251.
25. Martınez-Barrera, G., Menchaca-Campos, C., HernandezLopez, S., Vigueras-Santiago, E., & Brostow, W., (2006). Concrete reinforced with irradiated nylon fibers. *Journal of Materials Research, 21*, 484–491.
26. Martınez-Barrera, G., Menchaca-Campos, C., Vigueras-Santiago, E., & Brostow, W., (2010). Post-irradiation effects on Nylon-fibers reinforced concretes. *E-Polymers, 42*, 1–13.
27. Naskara, A. K., Mukherjee, A. K., & Mukhopadhyay, R., (2004). Studies on tyre cords: Degradation of polyester due to fatigue. *Polymer Degradation and Stability, 83*, 173–180.
28. Neppalli, N., Marega, C., Marigo, A., Bajgai, M. P., Kim, H. Y., & Causin, V., (2010). Poly(epsilon-caprolactone). filled with electrospun nylon fibers: A model for a facile composite fabrication. *European Polymer Journal, 46,* 968–976.
29. Pang, J. W. C., & Fancey, K. S., (2009). The flexural stiffness characteristics of viscoelastically prestressed polymeric matrix composites. *Composites A, 40*, 784–790.
30. Rajesh, C., Unnikrishnan, G., Purushothaman, E., & Thomas, S., (2004). Cure characteristics and mechanical properties of short nylon fiber reinforced nitrile rubber composites. *Journal of Applied Polymer Science, 92*, 1023–1030.
31. Rangari, V. K., Shaik, M. Y., Mahfuz, H., & Jeelani, S., (2009). Fabrication and characterization of high strength nylon-6/Si_3N_4 polymer nanocomposite fibers. *Materials Science and Engineering A-Structural Materials Properties Microstructure and Processing, 500*, 92–97.
32. Robinson, I. M., & Robinson, J. M., (1994). The influence of fiber aspect ratio on the deformation of discontinuous fiber-reinforced composites. *Journal of Materials Science, 29*, 4663–4677.

33. Saad, A. L. G., & Younan, A. F., (1995). Rheological, mechanical, and electrical properties of natural rubber-white filler mixtures reinforced with nylon 6 short fibers. *Polymer Degradation and Stability, 50,* 133–140.
34. Saeed, K., Park, S. Y., Haider, S., & Baek, J. B., (2009). In situ polymerization of multi-walled carbon nanotube/nylon-6 nanocomposites and their electrospun nanofibers. *Nanoscale Research Letters, 4*, 39–46.
35. Seema, A., & Kutty, S. K. N., (2006). Effect of an epoxy-based bonding agent on the cure characteristics and mechanical properties of short-nylon-fiber-reinforced acrylonitrile-butadiene rubber composites. *Journal of Applied Polymer Science, 99*, 532–539.
36. Shi, H. F., Zhao, Y., Dong, X., He, C. C., Wang, D. J., & Xu, D. F., (2004). Transcrystalline morphology of nylon 6 on the surface of aramid fibers. *Polymer International, 53*, 1672–1676.
37. Subbiah, T., Bhat, G. S., Tock, R. W., Parameswaran, S., & Ramkumar, S. S., (2005). Electrospinning of nanofibers. *Journal of Applied Polymer Science, 96*, 557–569.
38. Tian, M., Gao, Y., Liu, Y., Liao, Y., Xu, R., Hedin, N. E., et al., (2007). Bis-GMA/ TEGDMA dental composites reinforced with electrospun nylon 6 nanocomposite nanofibers containing highly aligned fibrillar silicate single crystals. *Polymer, 48*, 2720–2728.
39. Wang, R. X., Tao, X. M., Wang, Y., Wang, G. F., & Shang, S. M., (2010). Microstructures and electrical conductance of silver nanocrystalline thin films on flexible polymer substrates. *Surface & Coatings Technology, 204*, 1206–1210.
40. Younan, A. F., Ismail, M. N., & Yehia, A. A., (1992). Reinforcement of natural rubber with nylon 6 short fibers. *Journal of Applied Polymer Science, 45*, 1967–1971.
41. Zhou, Z., Xiao, Z., Pan, W., Xie, Z., Luo, X., & Jin, L., (2003). Carbon-coated nylon-fiber-reinforced cement composites as an intrinsically smart concrete for damage assessment during dynamic loading. *Journal of Materials Science & Technology, 19*, 583–586.

Chapter 8

Polyester Fiber-Reinforced Polymeric Materials

CONTENTS

8.1. Introduction 238
8.2. Synthesis and Basic Properties of Polyester Fibers 240
8.3. Polyester Fiber-Reinforced Polymer Composites 246
References 251

8.1. INTRODUCTION

The following definitions are being provided by the FTC (The Federal Trade Commission) for the polyester fiber: A fiber which contains a fiber-forming substance which is any long-chain synthetic polymer made up of a minimum of 85% by weight of an ester of a relieved aromatic carboxylic acid, with but not limited to substituted para-substituted hydroxy-benzoate units, p(-R-O-CO-C_6H_4-O-)x and terephthalic units, p(-R-O-CO-C_6H_4-CO-O-)x. On two polymers, the most general polyester fibers are based, namely poly(ethylene naphthalate) (PEN) and poly(ethylene terephthalate) (PET).

In the year of 1996, globally about a 24.1 million metric tons of industrial fibers were formed. In the production of PET fibers (PET staple 4% and PET filament 9%) the core volume gain took place (Froehlich, 1997). Through the demand for container and fiber resin, the motivation for this growth was taken (such as., water bottles and beverages). 75% of the whole PET production is generally directed on the road to fiber manufacturing. The world's three largest polyester producers are Eastman, Dupont, and Hoechst (Kim et al., 2007).

Every year the production of PET fiber in the world is growing and in 2001 reached an amount of around 20 million tons a year. About one decade ago the dramatic growth in PET fiber production was foreseen. Polyester has great resilience and strength, but its cost is lesser than that of rayon (Harris, 1996; Harrison, 1997).

In Dusseldorf, Germany, after 2 years, the worldwide leading plastic resin manufacturers are collected at the "K" plastic show organized to show their product range and new growths. The updated data and the market trends were shown in the 2007 event for the synthetic fiber production the general raw material and thermoplastic resins. In 2006 as per Koslowski (2007) the world manufacturing of plastics has touched about 245 million tons and 15% of them is about the production of synthetic fibers. As stated by Koslowski, the international production of PET capacity has touched >45 million tons yearly and globally with 6.6 million tons production in 2007 the polypropylene (PP) has importance, whereas in 2006 for PA resins the international market has been projected with 6.6 million tons (Wang et al., 2000; Silva et al., 2005).

In Table 8.1, we can see the developed poly(ethylene terephthalate) market growth between 2002 and 2008 (Li et al., 2006). The common names of PET products are: Tergal, Diolen, Dacron, Trevirafibers, and Terylene;

Cleartuf, Polyclear bottle resins, and Eastman; Mylarfilms, Melinex, and Hostaphan; and Ertalyte, Arnite, Impet, Valox injection-molding resins, and Rynite. The 18% of the world's total polymer production is done by the polyester industry and is presently after PP and polyethylene (PE) it is the third.

Through the recycled products now a well-established industry's branch yields PET fibers (Thiele, 2007). Total 3 to 5% of waste occurs in the manufacturing of PET fiber. After exposing to nature for a long time, the PET does not degrade. The wastes of PET are also quite useful because the PET is derived from petroleum and must be recycled because of ecological demands and economic reasons.

Two techniques are involved in the recycling of PET (Ambrosio et al., 1996; Chaouch et al., 2009):

- The mechanical recycling in which several unique properties polymer is being sustained or re-formed. Through mechanical shredding, the PET is transformed into flakes. Through the remelting, around 70% of the core portion, i.e., PET bottle flake is transformed into filaments and fibers. By using the remelting method along with an extruder, several properties of the PET wastes are enhanced. The properties of the final product decrease with degradation during remelting.
- The second method is chemical recycling, specifically back to the original raw materials, such as, DMT (dimethyl terephthalate) or PTA (purified terephthalic acid) and EG (ethylene glycol) in which the polymer structure is fully demolished (depolymerization), or in process intercedes, such as bis-b-hydroxyterephthalate.

Table 8.1. PET Market Size Per Year

Product Type	**2008 (Million Tons Per Annum)**	**2002 (Million Tons Per Annum)**
Textile-PET	39	20
Resin, bottle/A-PET	16	9
Special polyester	2.5	1
Film-PET	1.5	1.2
Total	**59**	**31.2**

According to the results by altering and carefully designing the parameters of the extruder, the properties of recycled PET wastes can be observed. Therefore, recycling costs are minimalized and additional degradation of the material can be avoided (Kim et al., 2002).

The PET fibers can be combined into polymer matrices ensuing in microscale polymer composites or can be straight merged with the melt of bi- or triphasic (hybrid) melts guiding the development of in situ-reinforced polymer composites (Ma et al., 2003).

8.2. SYNTHESIS AND BASIC PROPERTIES OF POLYESTER FIBERS

The PET is a general linear polyester as shown in Figure 8.1, polymerized either by straight esterification reaction among EG and terephthalic acid or ester exchange among the EG monomers and DMT (Fakirov, 2002). Both direct esterification and ester interchange processes are combined with polycondensation steps either continuously or batch-wise. As far as batch-wise systems are concerned, they require two reactors, one for polymerization and the other for esterification.

The solid-phase polycondensation is another method to produce PET in which a melt polycondensation is nonstop until an intrinsic viscosity of 1.0–1.4 dl/g is achieved. The polymer is cast into a solid firm at a definite point. Unless a target molecular weight is achieved, a precrystallization procedure is carried out through heating. After that for spinning the particulate polymer is melted as shown in Figure 8.2. This method is not much applicable for textile PET fibers, but it is employed for few industrial fibers.

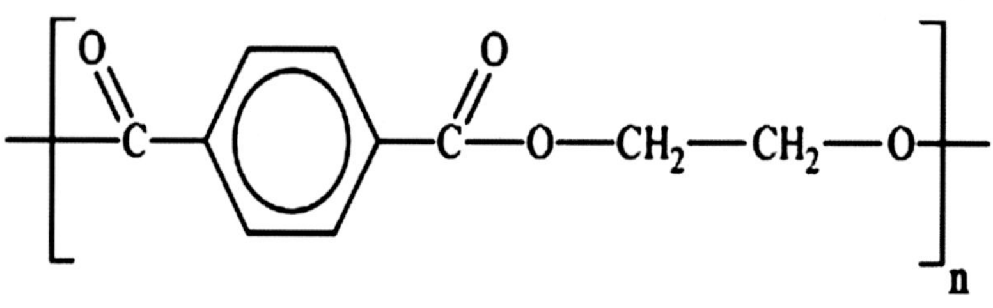

Figure 8.1. Chemical structure of PET.

Source: https://omnexus.specialchem.com/selection-guide/polyethylene-terephthalate-pet-plastic.

8.2.1. Fiber Manufacturing

Depending on its final uses, the degree of polymerization of PET can be controlled. PET employed for industrial fibers has a higher molecular weight, a higher degree of polymerization, and higher viscosity. The normal molecular weight ranges between 15,000 and 20,000 g/mol. Below normal extrusion temperature, it has a low shear viscosity of about 1000–3000 poise.

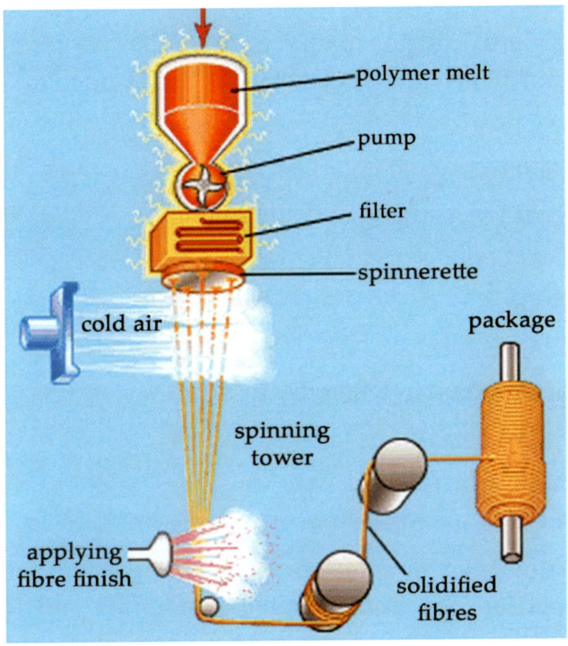

Figure 8.2. Polymer fiber melt spinning production process.

Source: https://www.textileflowchart.com/2015/08/flow-chart-of-melt-spinning-process.html.

At 265 C, the low molecular weight PET is spun, while at 300 C or more ultrahigh molecular weight PET is spun. Since it regulates the draw ratio, therefore, the degree of orientation is usually equivalent to the windup speeds in the spinning process (Huisman and Heuvel, 2003). At a wind-up speed of 10,000 m/min, the highest orientation together with the increase in productivity can be theoretically attained. Nevertheless, below accurate production condition effects like rejected fiber skin may show at wind-up speeds above 7000 m/min (Anoop et al., 2007). To yield uniform PET fiber surfaces, the process of drawing should be executed at a temperature above the typical unoriented polymer glass transition temperature (80–90 C).

Because the drawing process encourages extra orientation to products, the draw ratios (3:1–6:1) differ as per the end uses. As far as the higher tenacities are concerned the higher draw ratios must be employed. As projected along with crystallinity, orientation, at the temperature range of 140–220 C can be developed during the drawing (Cook, 1968).

8.2.2. Basic Properties of Polyester Fibers

The following properties are the main reason for the success of PET products in the market (Shin-Ichiro and Masayoshi, 2008; Marzougui et al., 2009):

- Strong;
- Resistant to several chemicals;
- Resistant to shrinking and stretching;
- Mold resistant;
- Fast drying;
- Wrinkle-resistant;
- Resilient and crisp when dry or wet;
- Abrasion-resistant;
- Mechanical response;
- Retains heat-set pleats and crinkle.

Because of molecular orientation and yielding higher crystallinity, the mechanical properties like initial Young's modulus and tensile strength generally improve as a positive function of the degree of fiber stretch. Final extensibility, such as, at the break elongation is generally reduced. As molecular weight increases, it also increases the elongation, modulus, and tensile properties (Jeddi et al., 2007; Vizesi et al., 2008). Common mechanical and physical properties of PET fibers are given in Table 8.2 and fiber-mechanical responses are shown in Figure 8.3. Through the table, we can see that staple (curves A and B) and high tenacity filament show very high moduli and breaking strengths but have relatively low elongations at the break. Curve C shows that it has a much greater initial modulus than the consistent tenacity essential shown in curve D. Alternatively, the last one shows a greater elongation and tenacity. POY (Partially oriented yarn) and spun filament yarns, display very high elongation but low strength (curve E) (Brown and Ward, 1968).

Table 8.2. Physical Properties of Polyester Fibers

Property	Filament Yarn		Staple and Tow	
	Regular Tenacity[a]	**High Tenacity**[b]	**Regular Tenacity**[c]	**High Tenacity**[d]
Elastic recovery at 5% elongation (%)	88–93	90	75–85	75–85
Breaking elongation	24–50	10–20	35–60	17–40
Specific gravity	1.38	1.39	1.38	1.38
Breaking tenacity (N/tex)[e]	0.35–0.5	0.62–0.85	0.35–0.47	0.48–0.61
Initial modulus (N/tex)[f]	6.6–8.8	10.2–10.6	2.2–3.5	4.0–4.9
Melting temperature (°C)	258–263	258–263	258–263	258–263
Moisture regain[g] (%)	0.4	0.4	0.4	0.4

[a] High strength and tire cord, high modulus industrial yarns.

[b] For knit and woven fabrics, the textile filament yarns.

[c] A consistent staple for carpet yarn, 100% polyester fabrics, fiberfill, and blends with cellulosic wool or blends.

[d] At 22C and 65% RH in the air, the standard measurements are conducted.

[e] For industrial applications, high modulus staple, sewing thread, High strength, and cellulosic blends.

[f] At 65% RH and 21C, the equilibrium moisture content of the fibers.

[g] To alter N/tex to ge/den (multiply by 11.33).

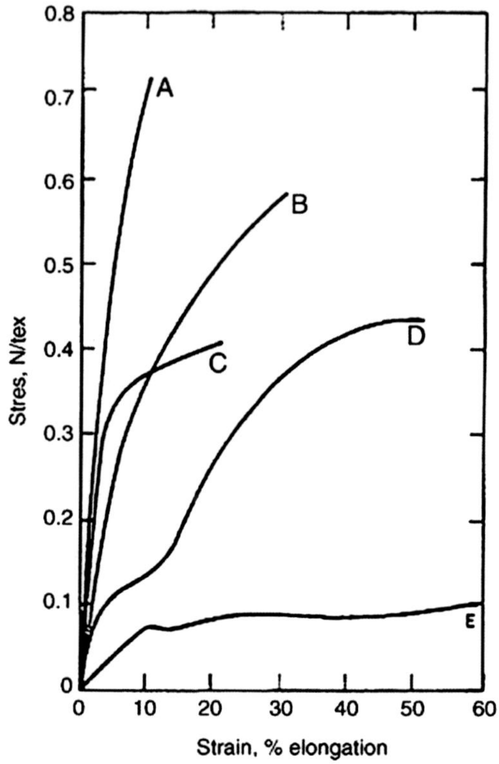

Figure 8.3. Tensile properties of PET fibers. (A) high tenacity filament; (B) high tenacity staple; (C) regular tenacity filament; (D) regular tenacity staple; (E) POY thread.

Source: https://www.tandfonline.com/doi/full/10.1080/23311916.2017.1362198.

Fatigue testing, such as revealing PET fiber to constant compression, so-called kink bands start to be initiated, and lastly subsequent in breaking of the kink band into a crack. In oriented PET, the slip (shear) bands through tensile testing have also been shown to seem. The compressive firmness of PET is higher than nylons (Hearle and Miraftab, 1995).

8.2.3. Fiber Viscoelastic Properties

According to few scientists, the high potential polyester fibers should be useful for applications where damage tolerance and high vibrational damping are usually needed. The ability of these fibers to absorb and shock loads or damp vibrations because of, in comparison, their polymer nature to the much stiffer carbon and glass fiber systems has a guide to studies about these loading cases. Usage of a polymer fiber is not bereaved of challenges concerning their viscoelastic nature; sensitivity to stress relaxation and creep hinters applications of polyester fibers where possible temperatures adjacent to glass transition, for long periods constant strains or long and high-term constant stress loading, are probable to happen (Kong et al., 2004; Demidov and Makarov, 2007).

House and Grant (1996) revealed that in marine structures, these systems could be combined, but their low stiffness must be taken into thoughtful attention. Davies et al. (2000) have researched polyester fiber's creep and relaxation behavior of anchorage lines and future a system for in-service strain measurements. After few years researchers proposed a nonlinear viscoelastic-viscoplastic model precisely explaining the creep behavior of cables made up of Diolen polyester (Chailleux and Davies, 2005). On the viscoelastic and crystallinity properties of various PET fibers, the role of manufacturing conditions has also been studied and it came to know that great fiber drawing rates guide to greater fiber crystallinity as confirmed by dynamic mechanical and X-ray diffraction analysis and fibers must be controlled through cooling to sustain high dynamic modulus data. With their viscoelastic behavior, the sewing polyester fiber properties have also been widely studied. These verdicts are quite significant as polyester fibers are currently often employed for sewing multiple layers of non-crimp fabrics for polymer composites (Ajiki and Postle, 2003; Dieval et al., 2004).

8.2.4. PEN Fibers

Poly(ethylene 2,6-naphthalate) has outstanding mechanical and thermal barrier and chemical resistance properties, but it owns a huge problem; it has very high viscosity for melt processing like injection molding, fiber spinning, and so on. It has been projected that the processability as per viscosity can be enhanced by adding other polymers like thermotropic liquid crystal polymers interested in the polymer matrix. Also, mixtures of PEN and PET have been appealing growing interest since they mix the greater properties of PEN along with the low price of PET (Cakmak et al., 1990; Ulcer and Cakmak, 1994). Because of transesterification through melt processing, the mixtures of these polyesters form arbitrary copolymers. Consequently, the glass transition temperature of PET/PEN rises linearly along with the volume part of PEN. It is, thus, probable to check properties associated with T_g by counting PEN to PET. The adding of PEN advances gas-barrier properties too. Instead, because of the higher stiffness of PEN low durability in repetitive bending can be predicted (Karayannidis et al., 2003).

Recently, through the electrospinning method, nanofibers of PEN has also been formed delivery very good results regarding attained fiber structure and diameter (1–2 mm). The least PEN nanofiber diameter ever described was recently attained through the exotic method of CO_2 laser beam supersonic jet drawing approved under vacuum (Schoukens and Clerck, 2005).

8.3. POLYESTER FIBER-REINFORCED POLYMER COMPOSITES

8.3.1. Elastomer Composites

In the last 20 years, significant effort has been made in strengthening several kinds of elastomers along with short polyester fibers. The basis behindhand this is the cords t, yarns, and PET fibers have been broadly employed as supports for vehicle tires, particularly the high tenacity-high modulus ones. Their practice is overwhelmed through PET degradation mechanisms like aminolysis and hydrolysis, which can help to early failure, sometimes because of fatigue initiation mechanisms. Instead, the application of long or short polyester fibers for elastomer reinforcement would make simpler things. The reason behind this is that the short fibers are much efficient to integrate into the elastomer resin and subsequent composites are acquiescent to the typical rubber processing steps: calendering, extrusion, and the several kinds of molding operations. In this way high-volume yields become more economical; therefore, it is feasible. This is in dissimilarity and very much cost-efficient than the slower procedures essential for joining and placing constant fibers in the form of cords or yarns. The hindrance is an expense in reinforcing strength along with irregular polymer fibers, though they significantly outperform common particulate fillers like carbon black (Jager et al., 1995; Kim et al., 2009).

For both polyurethane (PUR) and styrene rubber matrices, the effect of numerous bonding agents along with exciting outcomes has also been studied. NR (Natural rubber) composites are also widely studied; Natural rubber is an exciting matrix to strengthen with short polyester fibers. It came to know that electron-beam radiation in the natural rubber matrix can efficiently support chemical graft polymerization to silica particles, ensuing in strength and greater tensile modulus for allyl methacrylate and silanes coupling agents. Moreover, it has been efficiently confirmed that PET short fibers can offer greater price-to-mechanical concert features as potential NR reinforcements to that of aramid fiber pulp and sliced aramid (Saleh and Jabarin, 2001).

8.3.2. Microfibrillar-Reinforced (MFR) PET Composites

In order to improve the mechanical properties of polymeric materials founded on them the blending polyolefins along with engineering plastics can show as a beneficial solution. The ingredients in several polymer mixtures are thermodynamically incompatible and immiscible regarding processing. Consequently, through processing a large variety of shapes of the discrete phase can be molded, e.g., ellipsoids or spheres, plates, or fibrils. It is quite known that the size and shape of the discrete phase impact the properties of the last polymer blend. An unsuited pair of polymers can be managed in a way that the discrete phase makes *in-situ* fibers. Through this method, improved mechanical properties can be attained. Few scientists have suggested blending thermotropic LCPs (liquid crystalline polymers) in thermoplastics along with extensive success. Nevertheless, for many engineering applications, liquid crystalline polymers are not cost-effective (Higashioji and Bhusdan, 2002).

In contrast to LCPs, the main challenge when using general thermoplastics is that their molecular chains be susceptible to relaxation during melt processing; thus, a great molecular location is unavoidable. With the purpose of tackle this challenge, a novel kind of dispensation methodology, the so-called MFC (microfibrillar-reinforced composite) idea was given in the early 1990s. Along with them, preparation of MFC contains the subsequent three basic steps (Nakayama et al., 2008):

- Melt mixing with extrusion of two unmixable polymers containing various melting temperatures T_m (mixing step);
- Cold drawing of the extrudate with the good location of the two phases (fibrillization step);
- At temperature among the T_m's thermal treatment of the two mixture partners (isotropization step).

Several MFC systems contain PE and PP matrices usually. Li et al. suggested that the viscosity ratio among PE and PET should remain < 1, to get in the PE matrix the micro fibrillation of PET through slit-die extrusion. Li et al. used the vital work of the fracture toughness method to authenticate the hot stretch ratio in PET/PE MFCs and came out with the best relationship among them. Obviously, in the microfibrillar morphology build-up, the compatibilization of the polymer mixture ingredients also plays a vital part. It is coming to know that extruded 50/50 PET/HDPE blends by adding 7 wt.% of ethylene/methacrylic acid copolymer, outcomes in enhanced fracture toughness and ductility because of crack bridging through PET microfibers (Figure 8.4) (Suzuki and Yamada, 2010).

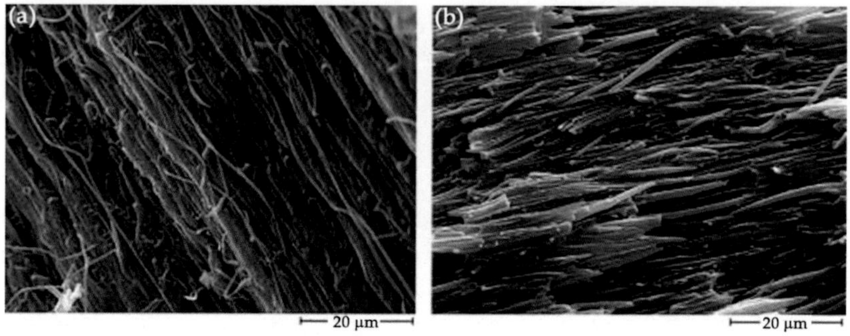

Figure 8.4. Surfaces of compression-molded PET/PP/E-GMA mixtures with an MFC structure in different ratios (by wt.%). (a) 40/60/0; and (b) 40/54/6. PP was selectively dissolved leaving PET fibrils are visible.

Source: https://www.sciencedirect.com/science/article/abs/pii/S0266353804001666.

PP/recycled PET MFCs have also been shown to display reinforcement and fibrillation effects when compatibilized with ethylene-glycidyl methacrylate throughout extrusion and then firmness molded later an orientation-drawing procedure (Hendricks, 2008).

8.3.3. Composites

As explained in the previous section, the greater action of polyester fibers in polymer composites regarding energy absorption, particularly below dynamic loading, such as impact was primarily recognized. The effort was made to combine these features into great performance graphite-epoxy/fiber matrix composites. Occasionally these composites are inserted along with one or more midway layers of PET fabrics, a method called "hybridization" leading to improved impact resistance. Researchers have stated that PET fiber mat when employed as an intermediate material can advance the crash characteristics of carbon fiber epoxy tubes (Hendricks, 2008). For high-performance carbon epoxy composites, the uniform PET films can be employed as interleaves to yield a similar toughening effect. The high fracture toughness can also be exhibited by short PET fiber-reinforced epoxy composites (1 wt.% fiber content) to improve adhesion. According to this, an almost 80% development was attained in K_{IC} determination done through quasi-static three-point bending tests (Figure 8.5) (Sawada et al., 1991).

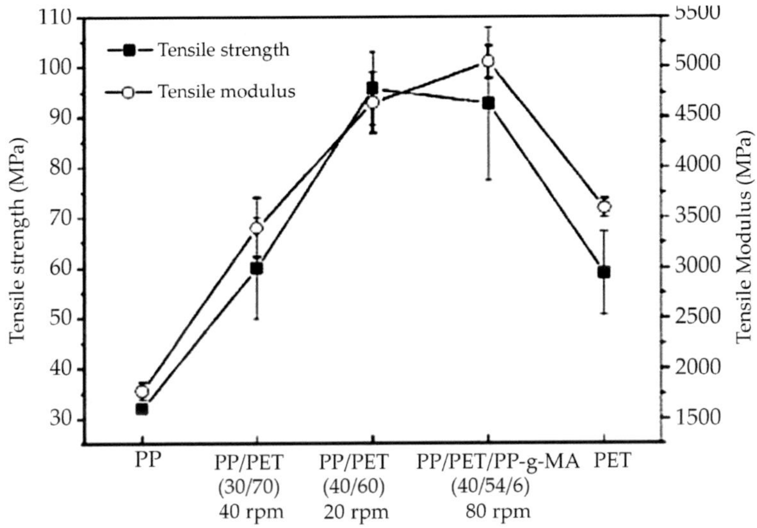

Figure 8.5. *Tensile properties of PET/PP microfibrillar composites.*

Source: https://onlinelibrary.wiley.com/doi/abs/10.1002/pen.21435.

About polyester fiber reinforcements, the thermoplastic matrices are also generally examined; the last being mixed as fillers previous to compounding. Scientists have discussed the effects of adding PET and nylon 66 short fibers, equally in unsized and sized states, on the dynamic and thermal properties of PP (Kikuchi, 1996). They came to know that 20 wt.% fiber content varies meaningfully in the PP matrix, and packing modulus E^0 crystallization kinetics. It was also been verified by Radhakrishnan and Saujanya on crystallinity (Carduner et al., 1989).

8.3.4. PET Nanocomposites

For nanocomposites due to the advancement in nanotechnology as stated in the introduction, has given motivation for manipulating the excellent properties of PET for a potential matrix. In the past decade, the commercial accessibility of CNFs (carbon nanofibers) directed scientists to examine the characteristics of systems of numerous polymers reinforced with carbon nanofibers. Because of the intrinsic low dispersibility of the carbon nanofibers, this is just a problematic operation into the polymer matrix. Cultured compounding approaches like high shear mixing in the melt, ball-milling, and extrusion by employing melt spinning and twin-screw extruders are now general techniques to formulate nanocomposites along with polymer matrices. For several kinds of CNFs, the PET has also been used as a matrix (Figure 8.6) (Ibarra et al., 1988).

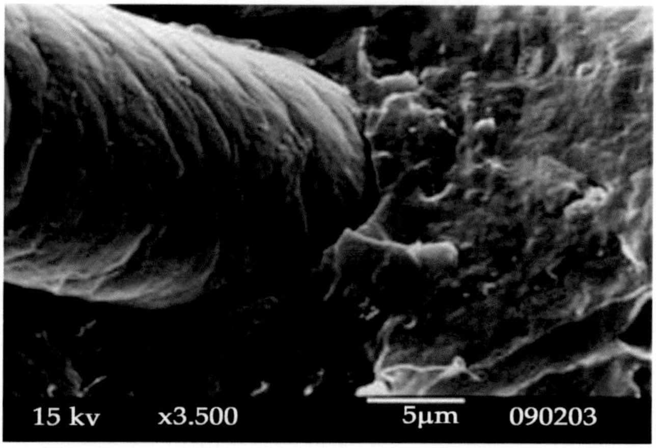

Figure 8.6. *A fiber of polyester pulled out from PP matrix displaying characteristic kink bands.*

Source: https://www.sciencedirect.com/science/article/abs/pii/S003238610000152X.

REFERENCES

1. Ajiki, I., & Postle, R., (2003). Viscoelastic properties of threads before and after sewing. *International Journal of Clothing Science and Technology, 15*(1), 16–27.
2. Ambrosio, L., Netti, P. A., Iannace, S., Huang, S. J., & Nicolais, L., (1996). Composite hydrogels for intervertebral disc prostheses. *Journal of Materials Science: Materials in Medicine, 7*(5), 251–254.
3. Anoop, A. K., Agarwal, U. S., & Rani, J., (2007). Carbon nanotubes reinforced PET nanocomposite by melt-compounding. *Journal of Applied Polymer Science, 104*(5), 3090–3095.
4. Brown, N., & Ward, I. M., (1968). Deformation bands in oriented polyethylene terephthalate. *Philosophical Magazine, 17*(149), 961–981.
5. Cakmak, M., Wang, Y. D., & Simhambhatla, M., (1990). Processing characteristics, structure development, and properties of uni and biaxially stretched poly(ethylene 2,6 naphthalate) (PEN) films. *Polymer Engineering and Science, 30*(12), 721–733.
6. Carduner, K. R., Paputa, P. M. C., Carter, R. O. III., & Killgoar, P. C. Jr., (1989). An infrared spectroscopic study of polyethylene terephthalate degradation in polyester fiber/nitrile rubber composites. *Polymer Degradation and Stability, 26*(1), 1–10.
7. Chailleux, E., & Davies, P., (2005). A non-linear viscoelastic viscoplastic model for the behavior of polyester fibers. *Mechanics of Time-Dependent Materials, 9*(2, 3), 147–160.
8. Chaouch, W., Dieval, F., Chakfe, N., & Durand, B., (2009). Properties modification of PET vascular prostheses. *Journal of Physical Organic Chemistry, 22*(5), 550–558.
9. Cook, J. G., (1968). *Handbook of Textile Fibers: Man-Made Fibers* (4th edn., Vol. 4, pp. 358, 361). Woodhead Publishing Ltd., Cambridge, UK.
10. Davies, P., Huard, G., Grosjean, F., & Francois, M., (2000). Creep and relaxation of polyester mooring lines. *Proceedings of the Offshore Technology Conference* (pp. 1–12). Paper OTC 12176.
11. Demidov, A. V., & Makarov, A. G., (2007). System analysis of the viscoelasticity of polyester fibers. *Fiber Chemistry, 39*(1), 83–86.

12. Dieval, F., Mathieu, D., & Durand, B., (2004). Comparison and characterization of polyester fiber and microfiber structure by x-ray diffractometry and viscoelasticimetry. *Journal of the Textile Institute, 95*(1), 131–146.
13. Fakirov, S., (2002). *Handbook of Thermoplastic Polyesters* (pp. 1200–1223). Wiley-VCH Verlag GmbH, Weinheim, ISBN: 3-527-30113-5.
14. Froehlich, F. W., (1997). Restructuring, innovation see Akzo Nobel through difficult business environment. *International Fiber Journal, 12*(3), 4–15.
15. Harris, W. B., (1996). Is there a future for polyester investments outside Asia? *International Fiber Journal, 11*(6), 4–9.
16. Harrison, D., (1997). *Synthetic Fibers for Nonwovens Update* (Vol. 28, No. 6, pp. 32–39). Nonwovens Industry.
17. Hearle, J. W. S., & Miraftab, M., (1995). The flex fatigue of polyamide and polyester fibers. Part II: The development of damage under standard conditions. *Journal of Materials Science, 30*(4), 1661–1670.
18. Hendricks, N., (2008). Diolen high-tenacity polyester yarns provides solutions for many markets. *International Fiber Journal, 23*(4), 44, 45.
19. Higashioji, T., & Bhusdan, B., (2002). Creep and shrinkage behavior of improved ultra-thin polymeric film. *Journal of Applied Polymer Science, 84*(8), 1477–1498.
20. House, J. R., & Grant, I. D., (1996). Viscoelastic composite materials for noise reduction and damage tolerance. *Advanced Performance Materials, 3*(3, 4), 295–307.
21. Huisman, R., & Heuvel, H. M., (2003). The effect of spinning speed and drawing temperature on structure and properties of poly(ethylene terephthalate) yarns. *Journal of Applied Polymer Science, 37*(3), 595–616.
22. IbarraRueda, L., Chamorro, A. C., & Tabernero, R. M. C., (1988). Mechanics of short fibers in filled styrene-butadiene rubber (SBR) composites. *Polymer Composites, 9*(3), 198–203.
23. Jager, J., Juijn, J. A., Van, D. H. C. J. M., & Huijts, R. A., (1995). Poly(ethylene-2,6 naphthalene dicarboxylate) fiber for industrial applications. *Journal of Applied Polymer Science, 57*(12), 1429–1440.
24. Jeddi, A. A. A., Nosraty, H., Taheri, O. M. R., & Karimi, M. A., (2007). Comparative study of the tensile fatigue behavior of cotton—polyester

blended yarn by cyclic loading. *Journal of Elastomers and Plastics, 39(*2), 165–179.

25. Karayannidis, G. P., Papachristos, N., Bikiaris, D. N., & Papageorgiou, G. Z., (2003). Synthesis, crystallization and tensile properties of poly(ethylene terephthalate-co-2,6-naphthalate)s with low naphthalate units content. *Polymer, 44*(26), 7801–7808.
26. Kikuchi, N., (1996). *Tires Made of Short Fiber Reinforced Rubber* (Vol. 214, No. 3, pp. 31, 32). Rubber World.
27. Kim, J. Y., Park, H. S., & Kim, S. H., (2007). Multiwall-carbon-nanotube-reinforced poly(ethylene terephthalate) nanocomposites by melt compounding. *Journal of Applied Polymer Science, 103*(3), 1450–1457.
28. Kim, M. S., Kim, H. K., Byun, S. W., Jeong, S. H., Hong, Y. K., Jood, J. S., Song, K. T., et al., (2002). PET fabric/polypyrrole composite with high electrical conductivity for EMI shielding. *Synthetic Metals, 126*(2, 3), 233–239.
29. Kim, S. Y., Kim, S. H., Lee, S. H., & Youn, J. R., (2009). Internal structure and physical properties of thermotropic liquid crystal polymer/poly(ethylene 2,6-naphthalate) composite fibers. *Composites: Part A, 40*(5), 607–612.
30. Kong, H., Mouritz, A. P., & Paton, R., (2004). Tensile extension properties and deformation mechanisms of multiaxial non-crimp fabrics. *Composite Structures, 66*(1–4), 249–259.
31. Koslowski, H. J., (2007). Global market trends for synthetic fiber polymers. *Chemical Fibers International, 57(*6), 287–293.
32. Li, Z., Luo, G., Wei, F., & Huang, Y., (2006). Microstructure of carbon nanotubes/PET conductive composites fibers and their properties. *Composites Science and Technology, 66*(7, 8), 1022–1029.
33. Ma, H., Zeng, J., Realff, M. L., Kumar, S., & Schiraldi, D. A., (2003). Processing, structure, and properties of fibers from polyester/carbon nanofiber composites. *Composites Science and Technology, 63*(11), 1617–1628.
34. Marzougui, S., Abdessalem, S. B., & Sakli, F., (2009). Viscoelastic behavior of textile artificial ligaments. *Journal of Applied Science, 9*(15), 2794–2800.

35. Nakayama, A., Takahashi, R., Hamano, T., Yoshioka, T., & Tsuji, M., (2008). Morphological study on electrospun nanofibers of aromatic polyesters. *Senni Gakkaishi, 64*(1), 32–35.
36. Saleh, Y. S., & Jabarin, A., (2001). Glass transition and melting behavior of PET/ PEN Blends. *Journal of Applied Polymer Science, 81*(1), 11–22.
37. Sawada, S., Kamiyama, K., Ohcushi, S., & Yabuki, K., (1991). Degradation mechanisms of poly(ethylene terephthalate) tire yarn. *Journal of Applied Polymer Science, 42*(4), 1041–1048.
38. Schoukens, G., & Clerck, K. D., (2005). Thermal analysis and Raman spectroscopic studies of crystallization in poly(ethylene 2,6-naphthalate). *Polymer, 46*(3), 845–857.
39. Shin-Ichiro, I., & Masayoshi, I., (2008). Irreversible phase behavior of the ternary blends of poly(ethylene terephthalate)/ poly(ethylene-2,6-naphthalate)/poly(ethylene terephthalate-co-ethylene-2,6naphthalate). *Journal of Applied Polymer Science, 110*(3), 1814–1821.
40. Silva, D. A., Betioli, A. M., Gleize, P. J. P., Roman, H. R., Gomez, L. A., & Ribeiro, J. L., (2005). Degradation of recycled PET fibers in Portland cement-based materials. *Cement and Concrete Research, 35*(9), 1741–1746.
41. Suzuki, A., & Yamada, Y., (2010). Poly(ethylene-2, 6-naphthalate) nanofiber prepared by carbon dioxide laser supersonic drawing. *Journal of Applied Polymer Science, 116*(4), 1913–1919.
42. Thiele, U. K., (2007). Polyester bottle resins production. In: *Processing, Properties and Recycling* (p. 259 ff). PET planet Publisher GmbH, Heidelberg, Germany, ISBN: 978-39807497-4-9.
43. Ulcer, Y., & Cakmak, M., (1994). Hierarchical structural gradients in injection-molded poly(ethylene naphthalene-2,6-dicarboxylate) parts. *Polymer, 35*(26), 5651–5671.
44. Vizesi, F., Jones, C., Lotz, N., Gianoutsos, M., & Walsh, W. R., (2008). Stress relaxation and creep: Viscoelastic properties of common suture materials used for flexor tendon repair. *The Journal of Hand Surgery, 33*(2), 241–246.
45. Wang, Y., Wu, H. C., & Li, V. C., (2000). Concrete reinforcement with recycled fibers. *Journal of Materials in Civil Engineering, 12,* 314–319.

INDEX

A

Abrasion resistance 60, 220
Acrylic polymers 56
Acrylonitrile butadiene rubber (NBR) 224
Aerodynamic 148
Aircraft application 187
Airframe structures 145
Alkaline environments 3
Alternative approach 230
Aluminum 174
Analytical expression 173, 183
Anhydrides 160
Anisotropic 218
Anisotropic behavior 204
Anisotropic thermal 203
Aramid fibers 104
Arbitrary radius 6
Aromatic polyamide 104

B

Basalt 27, 28, 29, 30, 39
Biochemistry 70
Biocomposites 92
Biosynthesis 72
Burning process 2

C

Caoutchouc 69
Carbon fibers 160, 162, 164, 165, 166, 167, 169, 170, 171, 177, 179, 180, 181, 182, 183, 185, 187, 190, 192, 193, 194, 195
Carbon fiber-strengthened polymer 186
Carboxylic acid 238
Cellulose 68
Cellulose molecules 68
Cellulosic fibers 4
Circular glass fibers 134
Circumference 94
Civil engineering materials 83
Community 2
Composite-covered transmission 187
Composite materials 2, 12, 14, 16, 22, 27, 35, 36, 82, 85, 86, 92

Composite plastics 86
Construction 2, 3, 4, 13, 18, 19, 21, 22, 32, 33, 36, 39, 40, 42, 43
construction industry 2, 13
Construction materials 33, 36, 40, 42
Corrosion resistance 186
Cresol-based polymers 64
Crystalline 66, 72, 74
Crystallinity 10, 67, 242, 245, 249
Crystallites 200
Crystallization 223, 229
Cupric hydroxide 69

D

Diazomethane 165
Diepoxides 61
Difunctional acids 60

E

Elastic beam 6
Electrical-glass 129, 130, 135, 136, 139, 143
Electrical-glass fibers 129, 139, 143
Electrical-glass in alkaline 139
Electrical insulation 53, 55, 65
Electrochemical treatment 166
Electrolysis 167
Electroplating 166, 179
Electrospinning 246
Energy-consuming 4, 20
Equilibrium distance 162
Ethylene terephthalate 238, 252, 253, 254
Exfoliate clay 231

F

Federal Trade Commission (FTC) 25
Fiber-forming material 8
Fiberglass 99, 100, 103, 117, 122, 123
Fiber material 3
Fiber-matrix interfacial 206
Fiber orientation 203, 204, 205, 211
Fiber production 221
Fiber-reinforced plastic (FRP) 82
Fiber-reinforced polymers 3, 82
Fiber-reinforced thermoplastics 201
Fiber reinforcement 86, 96, 103
Fiber reinforcing 99
Fiber spinning 7
Fiber spinning techniques 7
Fiber-strengthened composites 126, 128, 140, 141, 144
Fiber-strengthened polymers 127, 142
Fiber surface 207, 208, 209
Fibrous materials 4
Flat fabric 94
Flexibility 5, 6, 7, 8
Flexural rigidity 6
Fluids 67
Formaldehyde 62, 63, 64, 65, 74, 77, 78
Frequently phenolic 62

G

Gaseous mixtures 210
Geometric 96, 97, 98, 100, 106
Geometric shapes 96, 97, 98, 100, 106

Glass fiber production 131
Glass fibers 130, 132, 139, 153, 154
Glass-reinforced composition 89
Glass-reinforced fiber metal laminate (GLARE) 145
Glass-reinforced polymers 102
Golf clubs 186

H

High-density polyethylene fibers (HDPE) 19
High-pressure processes 54
Homopolymerization 59
Human operator 99
Hydrocarbon group 48
Hydrolysis 246
Hydroxy-benzoate 238
Hydroxy-groups 66, 68

I

Immiscible polymers possessing 230
Inadequate 2, 24, 31
Industrial production 53
Influence 222, 223, 226, 234
Inorganic 219, 223
Interfacial adhesion 160, 162, 167, 171, 179, 193, 222, 226
Interfacial shear strength (IFSS) 208
Interlaminar shear strength (ILSS) 171
Interlock equipment 95
International market 238
Intrinsic viscosity 240

K

Kinetics 52

L

Landfill space 2, 31
Linear elastic fracture mechanics (LEFM) 173
Linear polyamide 219
Liquid 7, 8, 9, 26, 228
Liquid polymeric matrix 162

M

Macromolecular 70
Macromolecules 50, 52, 60, 68, 73, 74, 76, 79
Macro synthetic fibers 18, 19
Manufacturing cost 4
Manufacturing process 162, 186
Matrix material 218, 219, 229
Matrix polymer 209
Matrix resins 160, 162, 166, 186
Mechanical behavior 160
Mechanical capability 129, 142, 144
Mechanical properties 242, 247
Metabolic processes 72
Metastyrene 50, 58
Methanoic 69
Methylene 220, 221
Methyl group 56, 72
Micrograph 223, 228, 229, 231
Microstructures 200
Microvoid 207
Microwave 148, 154
Modulus 199, 202
Moisture 54, 55, 132, 141, 144
Molecular 240, 241, 242, 247
Molecular mass 220
Molecular structures 199
Molecules 10
Morphology 67, 77, 218, 230, 232, 233, 235

Multilayer fabrics 95
Multilayer interlock braid 95

N

Natural environment 2
Nonstructural 2
Numerous manufacturing 161
Nylon stockings 61

O

Optimum physical properties 207
Organic liquid 49
Organic textile fibers 198
Oxidation 162, 163, 164, 165, 166, 176, 177, 189, 190, 191, 195
Oxidative nature 160
Oxide functional groups 178, 183
Oxyfluorination 167, 168
Oxygen-comprising fluorination 167

P

Particular technology 227
Performance 218, 219, 221, 232
Peroxide 54
Phenomenal 223
Phosphorus 48
Physical properties 136, 140, 141, 142, 144
Plasma treatment 162, 169
Plasticizers 112
Plastic material 53
Plastic waste 106
Polyacrylate 132
Polyamide (PA) 25
Polycondensation 86, 240
Polyester 238, 239, 240, 243, 244, 245, 246, 248, 249, 250, 251, 252, 253
Polyester resin 186
Polyethylene (PE) 3
Polyethylene terephthalate fibers (PET) 19
Polyethylene terephthalate (PET) 224
Polyfunctional 66
Polymer 48, 49, 50, 51, 53, 54, 55, 56, 57, 58, 59, 60, 63, 64, 65, 66, 68, 69, 70, 72, 77, 80
Polymer composites 88, 89, 114, 115, 119, 120, 121, 122, 123
Polymer-fabric layered preform 139
Polymer glass transition 241
Polymeric composites 160
Polymeric matrix 132
Polymerism 49
Polymerization 86, 90, 220, 221, 228, 229, 235, 240, 241, 246
Polymer matrices 198
Polymethylmethacrylate (PMMA) 224
Polypropylene (PP) 99
Polyurethane (PUR) 132
Polyvinyl acetate 132
Potential matrix 249
Precrystallization 240
Probability 225
Profound function 6
Propylene 55, 56

R

Radiofrequency 169
Raw materials 130
Reinforced thermoplastic resins 89
Reinforced thermosetting 89

Reinforcement 2, 3, 4, 13, 14, 16, 18, 19, 20, 21, 22, 24, 27, 29, 31, 33, 34, 38, 41
Reinforcement styles 83
Reinforcing fiber 93, 102
Resin transfer molding (RTM) 93, 140
Rigidity 57
Rubbery fillers 58

S

Sandwich structures 145
Sebacic acid 220
Semicrystalline polymers 229
Shock resistance 148
Silkworms 7
Starch 69, 75, 76
Strengthening material 139
Stress-strain behavior 112
Stretch flexible 112
Styrene 49, 50, 57, 58, 59, 74, 78, 79
Substantial energy 2
Surface free energy 201, 208
Surface oxygen 163, 164
Surfboards 148
Sustainable environment 85, 86
Synthetic fibers 3, 10, 13, 18, 19, 38, 43, 199
Synthetic polymers 86

T

Technical application 68
Tensile strength 206
Textile glass fibers 103
Textile manufacturing techniques 96
Thermoplastic 53, 65, 66, 72
Thermoplastic-based 198
Thermoplastic materials 140
Thermoplastic polymers 141
Thermoplastics 51, 53
Thermotropic liquid crystal 245, 253
Thickness 133, 142, 143
Timber 91
Titanium dioxide 220
Titanium tetrachloride 54
Toughness 91, 110
Transesterification 245
Transfer medium 203
Transformation 221
Transparent material 57
Triangular 134, 135, 150
Triethylaluminum 54
Trifunctional hydroxy 66
Triggering exclusive 127
Trimethylolpropane 66

U

Ultraviolet radiation 144
Unidirectional (UD) 228

V

Vacuum-assisted resin transfer molding (VARTM) 101
Vinyl ester (VE) 82
Viscosity 207, 209
Volume fraction 203, 206

W

Waals adsorption force 162
Waals attraction 160
Wood-plastic composites (WPC) 92
Worldwide leading plastic 238

Y

Yield metal matrix 166

Z

Ziegler-Natta catalyst 54
Ziegler processes 54

FSC
www.fsc.org
MIX
Papier | Fördert gute Waldnutzung
FSC® C083411

Kolibri 360 GmbH
Ferdinand-Jühlke-Straße 7
99095 Erfurt, Deutschland
produktsicherheit@kolibri360.de